Person, Thing, Robot

Person, Thing, Robot

A Moral and Legal Ontology for the 21st Century and Beyond

David J. Gunkel

The MIT Press
Cambridge, Massachusetts
London, England

The MIT Press would like to thank the anonymous peer reviewers who provided comments on drafts of this book. The generous work of academic experts is essential for establishing the authority and quality of our publications. We acknowledge with gratitude the contributions of these otherwise uncredited readers.

This book was set in ITC Stone Serif Std and ITC Stone Sans Std by New Best-set Typesetters Ltd. Printed and bound in the United States of America.

Library of Congress Cataloging-in-Publication Data

Names: Gunkel, David J., author.
Title: Person, thing, robot : a moral and legal ontology for the 21st century and beyond / David J. Gunkel.
Description: Cambridge, Massachusetts : The MIT Press, [2023] | Includes bibliographical references and index.
Identifiers: LCCN 2022033533 (print) | LCCN 2022033534 (ebook) | ISBN 9780262546157 (paperback) | ISBN 9780262375238 (epub) | ISBN 9780262375221 (pdf)
Subjects: LCSH: Robots—Moral and ethical aspects. | Androids—Moral and ethical aspects.
Classification: LCC TJ211.28 .G86 2023 (print) | LCC TJ211.28 (ebook) | DDC 629.8/92—dc23/eng/20221026
LC record available at https://lccn.loc.gov/2022033533
LC ebook record available at https://lccn.loc.gov/2022033534

10 9 8 7 6 5 4 3 2 1

Contents

Preface

Robots are a curious sort of thing. On the one hand, they are designed and manufactured technological artifacts. They are things. And like any of the other things that we encounter and use each and every day, they are objects with instrumental value. Yet on the other hand, these things are not quite like other things. They seem to have social presence, they are able to talk and interact with us, and many are designed to mimic or simulate the capabilities and behaviors that are commonly associated with human or animal intelligence. Robots therefore invite and encourage zoomorphism, anthropomorphism, and even personification.

So are robots things, technological objects that we can use or even abuse as we decide and see fit? Or is it the case that robots can or even should be something like a person—that is, another subject who would need to be recognized as a kind of socially significant other with some claim on us? These questions, which have been a staple in science fiction since the moment the robot stepped foot on the stage of history—quite literally in this case as the word *robot* was initially the product of a 1920 stage play by Czech playwright Karel Čapek—are no longer a matter of fictional speculation. They are science fact and represent a very real legal and philosophical dilemma.

Resolving this seems pretty simple. All that is needed is to assemble the facts and evidence, develop a convincing case, and then decide whether to categorize robots as one or the other. This is not just good reasoning: it's the law. In fact, the binary distinction separating who is a person from what is a thing has been the ruling conceptual opposition in both moral philosophy and jurisprudence for close to two thousand years. When the Roman jurist Gaius (130–180 CE), in a treatise he titled *Institutes*, explained that law involved two kinds of entities, either persons or things, he instituted a fundamental ontological division that has been definitive of Western (but not just Western) moral and legal systems. In the face of another—another human being, a nonhuman animal, a tree, an extraterrestrial, a robot, and so on—the first and perhaps most important question that must be addressed and resolved is "What is it?" Is it another *subject* similar to myself, to whom I would be obligated? Or is it just an *object* that can be taken up, possessed, and used without any further consideration or concern?

Consequently, all that is needed is to decide whether robots are things or persons. It sounds easy enough. But as detailed in the course of this book, it is much easier said than done. In fact, the robot does not quite fit in or easily accommodate itself to either category. Being neither an objectivized instrument that is a means to an end nor another kind of socially significant subject, the robot resists and confounds efforts at both reification and personification. It therefore frustrates and complicates the prevailing order—the mutually exclusive either/or—that has helped us make sense of ourselves and others by distinguishing who is to be recognized as a moral and legal subject from what remains a mere object or thing.

But this is not just about robots. It is ultimately about us. It is about the moral and legal institutions that human beings have fabricated to make sense of all that is. It therefore is about and concerns the fate of a myriad of *others* who we live alongside and that dwell with us on this exceptional and fragile planet. What is seen reflected in the face or faceplate of the robot is the fact that the existing moral and legal ontology is already broken or at least straining against its own limitations. And what is needed in response to this dysfunction is not some forceful reassertion of more of the same but a significantly reformulated moral and legal ontology that can scale to the challenges of the twenty-first century and beyond. Confronting and responding to this will undoubtedly be as terrifying and exhilarating as any of the robot uprisings that have been imagined in science fiction, because getting it right will require nothing less than a thorough rethinking of everything that we (and who is interpellated by or implicated in this first-person-plural pronoun will itself be an important component of what needs to be investigated) thought was right, natural, and beyond question.

What follows does not make any pretentions to providing a ready-made alternative. The existing ontology—the dichotomous model that divides persons from things—took hundreds if not thousands of years to be fully developed, codified, and instituted. It therefore seems reasonable to assume that a new moral and legal ontology—especially one that can respond to and take responsibility for the diversity of beings—will require the same kind of time, effort, and attention. For that reason, the objective of this book is more modest than what one might gather (or even expect) from a cursory reading of its title. Instead of delivering a brand-new, ready to use out of the box moral and legal ontology, it seeks to set the stage for that work, providing a structure and framework for what will be, and what needs to be, a deeply collaborative effort that invites and draws on the wide range and diversity of human knowledge, experience, and reflection. This book, then, is more of an invitation to work together in shaping a shared vision for the future than it is a determination and prescription for that future. It is the point of departure for what will need to be a shared journey and is not (not yet, at least) the destination.

In terms of my previous efforts, this book occupies the third position in what will now be the Machine Question trilogy. The first book in the series, *The Machine Question: Critical Perspectives on AI, Robots, and Ethics* (2012), sought to extend previous innovations in moral circle expansion by examining the terms and conditions by which machines of our own making would or could come to be considered moral agents and patients. The second, *Robot Rights* (2018), continued the analysis by examining whether and to what extent these artifacts either can or should have claims to moral status and legal recognition. And this third piece to the puzzle, *Person, Thing, Robot*, investigates how and why these artifacts already do not quite fit within the established order of things, challenging us to rethink and revise the existing moral and legal ontology.

Taken together, what these three books demonstrate is that robots, AI, and other seemingly social and intelligent artifacts are not just one more technological challenge to which we need to apply existing moral and legal thinking. They are and they do much more than that. They provide the opportunity for and provoke the need to reassess and reevaluate all aspects of what comprises and justifies these moral and legal affordances, questioning what is often taken for granted and thereby goes by without notice. This is the unique impact and significance of the machine question. It is a question not just of ethics applied to the exceptional opportunities and challenges of emerging technology. It is a question that concerns the very foundation and integrity of ethics itself.

Acknowledgments

The researching, writing, and publishing of a book is a communal activity. And, as has been the case with all of my previous literary efforts, this one has also benefited from productive interactions and dialogue with many coconspirators, colleagues, and critics.

The basic idea for the project initially came together in the wake of a debate that was organized and hosted by the Saint Thomas Aquinas Catholic Center of the University of Colorado, Boulder, in February 2020 (just as the COVID-19 pandemic was beginning to gain traction). The title for the event was "What is Personhood in the Age of AI?," and it pitted me against theologian Jordan Wales. That exchange and the transcript of our prepared remarks, which was eventually published in *AI & Society*, planted the seed for what has now grown and developed into this text. My gratitude to the Aquinas Center, for organizing the debate and graciously hosting the event; Jordan Wales, for his remarkable insights and engaging inquiries; the audience in Boulder, who had all kinds of interesting comments and important questions that they shared with us; and John-Stewart Gordon, who encouraged and helped facilitate publication.

The book began coming together in the fall of 2021 with the encouragement and support of my editor at the MIT Press, Philip Laughlin. Philip has now edited five of my books, and I am so very grateful to have had the fortune to work with him and the MIT Press. He not only provided crucial advice about the project and its approach, as he always does, but also sought out four incredibly insightful and useful reviewers. The reviewers were positive but devastatingly honest, and the text has truly benefited from their remarkable insights and critical feedback. I do not know who they are, but I am so very grateful for the work they did to help me think through some of the more difficult aspects of the material.

Work on the manuscript also benefited from podcast interviews and guest lecturing opportunities with colleagues from all over the world: Eileen M. Hunt, Bronwyn Williams, Kamil Mamak, Leigh M. Johnson, Rick A. Lee, Henrik Skaug Sætra, Vanessa Sinclair, Alex Martire, Ben Byford, John Danaher, Filipe Forattini, Joshua Smith, Tracey Follows, and Carson Littlefield. These conversations were not only fun and engaging but also provided a way to test-drive much of the material.

The revised manuscript, which was completed in April 2022, was graciously read by three friends and colleagues: Mark Coeckelbergh, Joshua Gellers, and Joshua Smith. I am as grateful for their enduring generosity and support as I am for the critical feedback and challenging questions that they always provide. I have learned so much from them, their own work, and the conversations we have had over these past several years. Editing of the manuscript was overseen by Kathleen A. Caruso and copyediting was provided by Melinda J. Rankin. Finally, none of this—the reading, the writing, and the publishing—would be conceivable without the support of my colleagues at Northern Illinois University, which has been my institutional home for over twenty years, and my family in Chicago. It is because of their enduring love and support that all of this is even possible.

1   Introduction

Ethics, in both theory and practice, is an exclusive undertaking. In confronting and dealing with others—other human beings, nonhuman animals, the natural environment, and technological artifacts—we inevitably make a decision that has the effect of dividing things into one of two types, arranging an exclusive distinction that separates those others who we are obligated to respect from what remain mere things. As eighteenth-century German philosopher Immanuel Kant (2012, 40) described it: "Beings whose existence rests not indeed on our will but on nature, if they are nonrational beings, still have only a relative worth, as means, and are therefore called *things*, whereas rational beings are called *persons*, because their nature already marks them out as ends in themselves, i.e., as something that may not be used merely as means, and hence to that extent limits all choice (and is an object of respect)."

But even if you've never heard of Kant, this just sounds intuitively correct. We go out into the world and deal with others, knowing there's a difference between other persons to whom we owe respect as ends in themselves and those things that are mere objects with instrumental value as a means to an end. As Roberto Esposito (2015, 1), who arguably wrote the book on this subject, explains: "If there is one assumption that seems to have organized human experience from its very beginnings it is that of a division between persons and things. No other principle is so deeply rooted in our perception and in our moral conscience as the conviction that we are not things—because things are the opposite of persons."

Not only does this distinction have deep roots, but it works.[1] And that's the problem. Because it works—and has worked so well—the line dividing things from persons is often taken for granted, assumed to be the natural state of affairs, and therefore is not questioned. It only stands out and becomes evident in those moments when it is challenged or even just momentarily violated. Consider, for example, what is now a rather common but still surprising social practice. Users of digital voice assistants, like Siri and Alexa, often find themselves saying "thank you" to the object. This is both weird and disorienting. We typically do not express gratitude to things or feel bad about not doing so. We use automobiles to travel around town without feeling the need to say "thank you" to the vehicle. But if we take a taxi or use a ride-sharing service, we will—or we think we should— say "thank you" to the driver of the vehicle, who we recognize as another person. Because digital voice assistants are things that talk like another person, we often (and rather unconsciously) respond to the object *as if* it were something other than a mere thing, a kind of someone to whom we feel obliged to say "thank you."

It is, of course, possible and entirely reasonable to explain and excuse these behaviors as mistakes. But what these "mistakes" reveal and make visible is that the line dividing person from thing (or, if you prefer more formalistic legal terminology, *persona* from *res*) is neither fixed nor stable. The boundary separating who is a person from what is a thing has been flexible, dynamic, and alterable. This is actually a good thing; it is a feature and not a bug. Ethics and law both innovate and advance by critically questioning their own exclusivity and accommodating many previously excluded or marginalized others, recognizing as persons what had previously been considered things and property—women, people of color, indigenous peoples, animals, the environment, and so on. And these critical challenges have often been spearheaded by innovative and forward-thinking efforts to introduce and advance a vindication of the rights of others—for example, Mary Wollstonecraft's protofeminist manifesto *A Vindication of the Rights of Women*, Thomas Taylor's vindictive but no less influential *A Vindication of the Rights of Brutes*, and now, it seems, a vindication of the rights of robots.

1.1 Robot Rights

The very notion of a vindication of the rights of robots sounds like something out of science fiction, and there is a good reason for this. Unlike artificial intelligence (AI), which originated in a scientific seminar held at Dartmouth College in the mid-1950s, robots are the product of fiction. And the idea of "robot rights" was already in play and operational from the moment the robot appeared on the stage of history. "The notion of robot rights," as Seo-Young Chu (2010, 215) has insightfully pointed out, "is as old as is the word 'robot' itself. Etymologically the word 'robot' comes from the Czech word *'robota,'* which means 'forced labor.' In Karel Čapek's 1920 play *R.U.R.*, which is widely credited with introducing the term 'robot,' a 'Humanity League' decries the exploitation of robot slaves—'they are to be dealt with like human beings,' one reformer declares—and the robots themselves eventually stage a massive revolt against their human makers." This scenario—the robot uprising and struggle for recognition—has become one of the most popular themes or leitmotifs in subsequent science fiction. Even if you are not familiar with Čapek's play, there's a good chance you already know the storyline.

So when we first hear or encounter the phrase *robot rights* or *rights for robots*, we can be excused if the expectation is for some kind of dramatic uprising as the machines either take up arms against their human oppressors or take to the streets in violent protest. But reality is fortunately much more sanguine, considerably less dramatic, and, in a word, *real*. This is not to say that the robot uprising will not happen. It will. In fact, it already has. It just does not conform to what has been scripted for us in science fiction. No robot armies marching through the streets, no heart-pounding chase sequences with spectacular explosions, no intense emotional encounters between the robots and their makers. None of that. Instead, it takes place and is already taking place in ways that are much more mundane and seemingly boring by comparison, as philosophers, legal scholars, and courts and legislatures advance proposals, publish documents, and deliberate costs and benefits. It's more C-SPAN than it is *Terminator*.

Consider the following: There has been an explosion of activity addressing the subject of robot rights in both academic research and popular media. In a survey of the existing scholarly literature, Jamie Harris and Jacy Reese Anthis (2021) found just under three hundred publications in circulation, with an exponential rise in activity over the past several years. There are books with provocative titles like *Robot Rights* (Gunkel 2018), *Rights for Robots* (Gellers 2020), *Artificial Life after Frankenstein* (Hunt Botting 2021), *Robotic Persons* (Smith 2021a), *We, the Robots?* (Chesterman 2021), and more. There are peer-reviewed research articles published in journals of ethics, artificial intelligence, and law. Here's just a sample of titles: "Legal

Personhood for Artificial Intelligence" (Jaynes 2020), "The Hard Problem of AI Rights" (Andreotta 2021), "Recognising Rights for Robots" (Bennett and Daly 2020), "Is it Time for Robot Rights?" (Müller 2021), and "Robots and Rights" (Schröder 2021).

In the popular press, one can find articles like the *New Yorker*'s "If Animals Have Rights, Should Robots?" (Heller 2016), "Robot Rights?" in the *Brown Political Review* (Lehman-Ludwig 2019), *Diginomica's* "Robot Rights—A Legal Necessity or Ethical Absurdity?" (Marko 2019), "Humans Keep Directing Abuse—Even Racism—at Robots" at Vox (Samuel 2019), and video explainers, like "Do Robots Deserve Rights?" from Kurzgesagt—In a Nutshell (2017) and an interview with animal rights innovator and Berggruen Prize winner Peter Singer in "Will Robots Have Rights in the Future?" from Big Think (2019).

But this subject is not something limited to academic curiosity and popular media speculation. "The robot rights argument," as James Dawes (2020, 592) points out, "has already begun—just Google the phrase." And if you do so, you will find that there are already a number of actual proposals in circulation and even a few legislative acts and judicial decisions already on the books. In May 2016, the Committee on Legal Affairs of the European Parliament—the legislative branch of the European Union—proposed that "sophisticated autonomous robots" be considered "electronic persons" with "specific rights and obligations" for the purposes of contending with the challenges of technological unemployment, tax policy, and legal liability.

In November 2020, the legislature of the Commonwealth of Pennsylvania passed a bill (Senate Bill 1199) that classifies autonomous delivery robots, or what the text of the act calls personal delivery devices (PDDs), as pedestrians in order to provide a legal framework for their deployment on city streets and sidewalks. Similar laws have been passed in a number of other jurisdictions, including the commonwealth of Virginia (Code of Virginia § 46.2–908.1:1), which provides the following stipulation: "a personal delivery device operating on a sidewalk or crosswalk shall have all the rights and responsibilities applicable to a pedestrian under the same circumstance."[2]

In 2021, South Africa and Australia recognized an artificial intelligence system as the inventor on a patent application. This outcome was the result of an international effort lead by legal scholar and lawyer Ryan Abbott. Since 2018, Abbott and his team at the Artificial Inventor Project have petitioned patent offices across the globe to recognize an AI system developed by Stephen Thaler, called DABUS (Device for the Autonomous Bootstrapping of Unified Sentience), as the sole inventor of a food container system. Previous filings in the EU, UK, and US had been denied—not because DABUS did not originate the product but because, under current law, only natural persons may be named as inventors on a patent application. While these decisions were being appealed, first South Africa's Companies and Intellectual Property Commission and then the Federal Court of Australia (in the decision regarding *Thaler v. Commissioner of Patents* [2021] FCA 879) found in favor of the applicant, becoming the first jurisdictions on planet Earth to recognize the claim of an AI to be legally recognized as an inventor on a patent application.

1.2   The Debate

Robot rights are not a matter for the future. They already matter here and now. Even if real-world circumstances and scenarios are seemingly less exciting and action-packed than the robot uprisings of science fiction, the issue of robot rights is the site of a dramatic and important conflict. And like any conflict, there are two sides or opposing forces.[3] On one side, there are what could be called the Critics. According to this group, the very idea of robots, AI applications, or other socially interactive machines being accorded anything approaching moral or legal status beyond that of a mere instrument or piece of property is not just wrong-headed thinking but also a dangerous development that should be severely curtailed, resisted, or interrupted before it even begins. In short: robots, AI systems, and other artifacts are just things and not persons. The other side—formed of what we might call, by way of contrast, the Advocates—recognizes that various technological systems and implementations might need some form of social recognition and/or legal protection and that entertaining this exigency is an important contribution to ongoing efforts to test, validate, and even revise the limits of our moral and legal systems. In short: robots, AI systems, and other artifacts can be people too (figure 1.1).[4]

The debate is polarizing, with one side opposing what the other promotes. One side, for instance, argues that robot rights open the opportunity for thinking about the limitations of existing moral and legal systems, thereby contributing to similar efforts to address the plight of previously

Advocates: Robots and AI might need some form of social recognition and protection; entertaining this exigency is an important component in our on-going efforts to test, validate, and even revise the limits of our moral and legal systems.

Critics: The very idea of robots or AI being accorded anything approaching moral or legal status is not just wrong-headed thinking but a dangerous development that should be severely curtailed, resisted, or shut down before it even begins.

Figure 1.1
The robot rights debate. Original image by the author.

excluded individuals and populations. The other side argues that focusing attention on what are human designed and manufactured artifacts actually distracts us from the more important and pressing moral, legal, and social matters that confront us, thereby risking further harm to already vulnerable populations. One side suggests that as robots and AI systems become increasingly capable, sentient, and maybe even conscious, we will need to consider their interests and well-being in a way that is no different from the consideration enjoyed by other persons, like human beings or even non-human animals. The other side argues that because robots with consciousness or sentience would need the protection of rights, it would be prudent to avoid ever making things like this to which we would feel the need to be obligated. One side proposes that addressing questions regarding robot rights and the legal standing of AI systems will help us resolve problems of liability and responsibility in a world where artifacts make (or at least seem to make) independent decisions. The other side asserts that doing so will only exacerbate existing problems with responsibility gaps, shell companies, and liability shields.

It is a heated contest, with both sides appearing to advance positions and arguments that (when one initially hears them) make sense. And like similar polarizing disagreements—think, for example, of other seemingly irresolvable moral or legal disputes, like debates about abortion or physician-assisted suicide—there is no clear winner. Both sides continue to heap up arguments and evidence in support of their position, but the basic terms and conditions of the conflict remain largely in place and essentially unchanged. For this reason, this book does not take sides in the existing conflict by advocating for one over and against the other, nor does it seek to mediate their differences and disputes. Instead, it deploys an altogether different strategy. It targets not the points of conflict nor the differences that separate the one from the other, but the common set of shared values and fundamental assumptions that both sides already endorse and must endorse in order to enter into conflict in the first place. And it does so in order to devise an alternative that can better respond to and take responsibility for the moral and legal opportunities and challenges that we confront in the face or the faceplate of robots.

1.3 Terminology

Before getting too far into things, it may be prudent to pause for a moment and define or at least characterize the term *robot*. As indicated earlier, the word originates in a work of fiction—specifically, a stage play titled *R.U.R.* or *Rossumovi Univerzální Roboti* (*Rossum's Universal Robots*), written by Karel Čapek. In Czech, as in several other Slavic languages, the word *robota* (or some variation thereof) denotes "servitude or labor," and *robot* was the word that Čapek used to name a class of manufactured, artificial servants. Since the publication of Čapek's play, robots have infiltrated the space of fiction. And some of the most memorable characters in twentieth- and twenty-first-century film and television have been robots: Robby the Robot, Astroboy or Tetsuwan Atomu, Data, R2-D2 and C-3PO, WALL-E and EVE, the Cylons, the replicants, and the Terminator.

When it comes to defining the term *robot*, science fiction actually plays a significant and influential role. In fact, much of what we know or think we know about robots comes not from actual encounters with the technology but from what we see and hear about in fiction. When you ask someone—especially someone who is not a roboticist—to define *robot*, chances are the answer that is provided will make reference to something found in a science fiction film, television program, or story. This does not only apply to or affect outsiders looking in. "Science fiction prototyping," as Brian David Johnson (2011) calls it, is rather widespread within the disciplines of AI and robotics, even if it is not always explicitly called out and recognized as such. As roboticists Bryan Adams, Cynthia Breazeal, Rodney Brooks, and Brian Scassellati (2000, 25) point out: "While scientific research usually takes credit as the inspiration for science fiction, in the case of AI and robotics, it is possible that fiction led the way for science."

So what in fact does the word *robot* designate? Even when one consults knowledgeable experts, there is little agreement when it comes to defining, characterizing, or even identifying what is (or what is not) a robot. In the book *Robot Futures*, Illah Nourbakhsh (2013, xiv) explains the problem this way: "Never ask a roboticist what a robot is. The answer changes too quickly. By the time researchers finish their most recent debate on what is and what isn't a robot, the frontier moves on as whole new interaction technologies are born."

Despite this equivocation, definitions are unavoidable and necessary. One widely cited source of a general, operational definition comes from George Bekey's *Autonomous Robots: From Biological Inspiration to Implementation and Control*: "In this book we define a robot as a machine that senses, thinks, and acts. Thus, a robot must have sensors, processing ability that emulates some aspects of cognition, and actuators" (Bekey 2005, 2). This "sense, think, act" or "sense, plan, act" (Arkin 1998, 131) paradigm has considerable traction in the literature—as evidenced by the fact that it constitutes and is called a *paradigm*.

This characterization of a robot is, as Bekey (2005, 2) explicitly recognizes, "very broad," encompassing a wide range of different kinds of technologies, artifacts, and devices. But it could be seen as being too broad insofar as it may be applied to all kinds of artifacts that exceed the conceptual limits of what many consider to be a robot. As John Jordan (2016, 37) notes, "The sense-think-act paradigm proves to be problematic for industrial robots: some observers contend that a robot needs to be able to move; otherwise, the Watson computer might qualify." The Nest thermostat provides another complicated case: "The Nest senses: movements, temperature, humidity, and light. It reasons: if there's no activity, nobody is home to need air conditioning. It acts: given the right sensor input, it autonomously shuts the furnace down. Fulfilling as it does the three conditions, is the Nest, therefore, a robot?" (37). And what about the seemingly common and mundane smartphone? According to Joanna Bryson and Alan Winfield (2017, 117), these devices could also be considered robots under this particular characterization. "Robots are artifacts that sense and act in the physical world in real time. By this definition, a smartphone is a (domestic) robot. It has not only microphones but also a variety of proprioceptive sensors that let it know when its orientation is changing or when it is falling."

Consequently, *robot* already allows for and encompasses a wide range of different concepts, entities, and characterizations. It therefore is already the site of a conversation and debate about technology and its social position and status, and we should not be too quick to close off the possibilities that this lexical diversity enables and makes available. As Andrea Bertolini (2013, 216) argues, "All attempts at providing an encompassing definition are a fruitless exercise: robotic applications are extremely diverse and more insight is gained by keeping them separate." Adopting this kind of approach—one that is tolerant of and can accommodate a range or an array of different connotations and aspects—allows for a term like *robot* to be more flexible and for the analysis to be more responsive to the diverse ways the word is actually utilized and applied across different texts, social contexts, research efforts, historical epochs, and so on.

This does not mean, however, that anything goes and that *robot* (or *AI*, which is often substituted for *robot*[5]) is whatever one wants or declares it to be. It means, rather, paying attention to how the term comes to be deployed, defined, and characterized in the scholarly, technical, and popular literature, including fiction; how the term's connotations shift over time, across different contexts, and even (at times) within the same text; and how these variations relate to and have an impact on the options, arguments, and debates concerning the moral and legal status of these technological artifacts.

1.4 Plan of Attack

The analysis at hand will commence by getting a handle on both *persons* and *things*. Doing so will involve excavating from the sediment of the history of philosophy and law not only the way that both concepts have been deployed and developed but also, and perhaps more importantly, how they have been distinguished from each other and how that difference has shaped the way that each term has come to be defined and operationalized. As Esposito (2015, 16) explains: "From time immemorial our civilization has been based on the most clear-cut division between persons and things. Persons are defined primarily by the fact that they are not things, and things by the fact that they are not persons."

Although the two categories of this mutually exclusive and totalizing conceptual order appear to be relatively stable, membership is not. Over time, many "things" that were once regarded as things—women, children, slaves, animals—have come to be recognized as persons and therefore admitted into the community of moral and legal subjects. What's currently up for debate, then, is whether robots, AI systems, and other artifacts belong solely and exclusively to the category of *thing*; whether it is possible now or in the future that these entities might cross the line and be recognized as persons, possessing rights and obligations; or whether we might be able to split the difference and formulate some kind of third alternative that is neither the one nor the other. And a good part of the analysis that follows consists of a detailed reading and thorough cost/benefit analysis of the various arguments that have been deployed in this domain by partisans on both sides of the debate. As with previous books, especially *Robot Rights* (Gunkel 2018), the objective of this undertaking is to document who is arguing what, to examine how the different arguments and variations have been asserted and formulated, and ultimately to figure out what it all means. In other words, the goal is to understand how all the piece of this complex puzzle fit together.

What is of primary importance in this undertaking is not what makes one side in the debate different from and/or opposed to the other. What is of interest is what both sides already agree upon and endorse in order to come into conflict to begin with. Despite their disagreements and often polar-opposite opinions, what is not up for debate or submitted to questioning is the fundamental ontological presupposition that distinguishes person from thing. This difference, which is asserted and operationalized by both the Critics and the Advocates as if it were some universally true and naturally justified determination that has persisted from the beginning of time, is specific and context dependent. It is the product of a particular cultural formation and philosophical tradition. Consequently, the real problem is not that one side is different from and opposed to the other. The problem is that both sides tacitly agree to one way of dividing up the world and then quibble about who or what is to be included or excluded from one category or the other. In other words, both sides agree to and play by the same set of rules—that is, a shared moral and legal ontology. But these rules (like all rules to all games) are arbitrary, alterable, and at least something that needs to be submitted to critical investigation and reappraisal.

The objective of this book, then, is to intervene in this conceptual order in such a way as to neither endorse one side or the other nor to mediate their differences via some kind of third alternative that would resolve the dispute. There is name for this kind of critical intervention: deconstruction.[6] The word *deconstruction*, despite initial perceptions, does not indicate "to take apart," "to un-construct," or "to disassemble." Despite this widespread and rather popular misconception, which has become something of an institutional (mal)practice in both popular media and academic circles, deconstruction is not negative. But to say that it is not negative does not mean that it is something positive. Instead, what deconstruction designates is a kind of thinking outside the box, what in Aristotelean logic would be called the *law of noncontradiction*, that exceeds the grasp of the existing conceptual order and its oppositional pairs—for example, construction/destruction, positive/negative, inside/outside, person/thing, and so on. But *how* this transpires and (maybe more importantly) *why* are questions of methodology.

1.5 Method of Analysis

With any conceptual opposition, the two opposing terms are not situated on a level playing field; one of the two already has the upper hand. In the person/thing dichotomy, for instance, it is *person* that occupies this privileged position. "The relation between" things and persons, as Esposito (2015, 17) explains, "is one of instrumental domination, in the sense that the role of things is to serve or at least to belong to persons. Since a thing is what belongs to a person, then whoever possesses things enjoys the status of personhood and can exert his or her mastery over them." Deconstruction of this or any of the other binary oppositions that organize systems of thinking—whether philosophical, legal, economic, political, or ethical—proceeds by way of a double gesture, or what has also been called a "double science" (Derrida 1981, 41).

1.5.1 Double Science

In a first move, we deliberately invert the two terms that make up the existing conceptual order. In the person/thing dichotomy, *person* occupies the position of privilege and has been granted dominance over *thing*. So we begin by flipping the script. This operation is quite literally a *revolutionary* gesture insofar as the existing order—an arrangement that is already an unequal and violent hierarchy—is inverted or overturned. "To overlook this phase of overturning," Derrida (1993, 141) explains, "is to forget the conflictual and subordinating structure of opposition."

But inversion, in and by itself, is not sufficient. It is only half the story. This is the reason that it is just a phase or first step. As Derrida points out, a conceptual inversion or revolutionary overturning—whether it be social, political, or philosophical—actually does little or nothing to destabilize the existing order or really change things. In merely exchanging the relative positions occupied by the two opposed terms, inversion still maintains and preserves the binary opposition in which and on which it operates—albeit in reverse order or upside down. This is precisely the problem that is dramatized and explored in science fiction with the proverbial robot uprising.

In rising up in revolution against their human makers, robots overturn the existing social order, replacing the dominance of human persons with robotic things and artificially intelligent machines. Revolution, then, just reverses the existing hierarchy and, in doing so, changes little or nothing. It may provide the opportunity for some good cinematic drama and action sequences, but it is no solution. This is because, as Derrida (1981, 41) knew and pointed out, mere revolutionary inversion still "resides within the closed field of these oppositions, thereby confirming it." To put it another way, if we stopped here, at this phase of overturning, siding with things over and against persons, then it would be hard to answer for or respond to the charge that this revolutionary effort amounts to little more than an antihumanism or depersonalization.

For this reason, deconstruction necessarily entails—and must proceed to—a second, postrevolutionary phase or operation. "We must," as Derrida (1981, 42) states, "also mark the interval between inversion, which brings low what was high, and the irruptive emergence of a new 'concept,' a concept that can no longer be, and never could be, included in the previous regime." Strictly speaking, this new concept is no concept whatsoever, for it always and already exceeds the system of oppositional logic that defines the conceptual order as well as the nonconceptual order with which the conceptual order has been articulated. This so-called new concept—the very naming of which requires either repurposing the resources of already existing words (what Derrida calls *paleonymy*) or inventing new ones (something called *neologism*)—occupies a position that is outside of or at the margins of a traditional, conceptual opposition or binary pair. And, as we will see, this new concept, this Thing that is neither a person nor a thing, can be identified with the name *robot*.

1.5.2 Raison d'être

But this abstract and schematic characterization of the double gesture of deconstruction does not answer the more basic question: Why? If, since Roman times, the conceptual distinction separating persons from things "has been reproduced in all modern codifications, becoming the presupposition that serves as the implicit ground for all other types of thought—for legal but also philosophical, economic, political, and ethical reasoning" (Esposito 2015, 2), then why would we ever mess with it? If the person/thing dichotomy has worked and continues to work, why bother questioning it at all? What works, works. Isn't that good enough? Actually, no. And we need to mess with it for several reasons:

1. Limitations.   Binary oppositions, although useful for categorizing things, restrict what is possible to know and to say about the world and our own experiences. This is because conceptual opposites push things toward mutually exclusive options. In this either/or mode, any phenomenon is assumed to be reducible to x or its opposite, *not-x*. In other words, we typically make sense of ourselves and our world by deploying sets of terminological differences or conceptual oppositions, like that which divides entities into the categories of persons and things. As Barbara Johnson (1987, 12) explains, the underlying logic of this way of thinking—that is to say, "if not absolute, then relative; if not objective then subjective; if you are not for something; you are against it"—is the principle of noncontradiction. This principle, or what is also called the *law of noncontradiction*, has been, at least since the time of Aristotle, one of the defining conditions—if not the defining condition—of human knowledge. As Paula Gottlieb (2019) explains: "According to Aristotle, first philosophy, or metaphysics, deals with ontology and first principles, of which the principle (or law) of noncontradiction is the firmest. Aristotle says that without the principle of noncontradiction we could not know anything that we do know."

Although this kind of exclusivity has a certain functionality and logical attraction—not to mention the fact that it has the status of being not just *a* law but *the* law—it is often criticized for not being entirely in touch with the complexity and exigency of facts on the ground. It is for this reason that we are generally critical of *false dichotomies*—the parsing of complex reality into simple either/or distinctions. And the person/thing dichotomy, as Anna Beckers and Gunther Teubner (2021, 13) point out in the context of AI and robot law, provides an almost perfect illustration of the problem, insofar as existing technology already seems to resist both reification, where the AI or robot is "just a tool" of human action, and personification, promoting the artifact to a moral or legal position that would be similar to that of a natural human person. There are, therefore, ontological and epistemological reasons to question the hegemony of rigid binary oppositions and the structural limitations that they impose.

In addition, and following from this, conceptual opposites arrange and exert power. The two items are not situated on a level playing field; one of the pair has already been determined to be the privileged term. "We are not," as Derrida (1981, 41) explains, "dealing with the peaceful coexistence of a *vis-à-vis*, but rather with a violent hierarchy. One of the two terms governs the other (axiologically, logically, etc.), or has the upper hand." Consequently, binary oppositions are not just descriptive or a matter of some neutral discursive difference; they are the site of real social, political, and moral power. Whoever gets to divide up the world into this versus that or us versus them has the power to direct and determine what is possible to think, say, and do. The promise of deconstruction is that it provides a potent mechanism for working our way out of the maze of oppositional pairs and dualisms, like person/thing, by which we have made sense of ourselves, our world, and others. And this is especially important for those individuals and communities who have been, for one reason or another, situated on the "wrong side" of these oppositional dualities—all those who, as Donna Haraway (1991) describes it, have been assigned to the unfortunate position of being the other of (Western) man.

2. Gridlock. The debate about how robots, AI systems, and other seemingly intelligent artifacts fit into the existing ontological categories of person or thing has produced a lot of activity, but the effort seems to be caught in what nineteenth-century German philosopher G. W. F. Hegel (2010, 202) had called a "bad infinity"—a seemingly endlessly repetition of the same arguments and disputes that do not make much progress on the issue. After several decades of work and robust discursive activity—activity that will be documented and analyzed in detail—we haven't got very far and find ourselves in a kind of stalemate or cul-de-sac. There are seemingly good and decent reasons that robots and AI systems cannot and should not be regarded as mere things. But there are just as many good arguments and evidence that say extending the status of person to these machines would be just as bad, if not worse.

So at this juncture, we find ourselves at something of an impasse. Each side in the debate—the Critics and the Advocates—continue to heap up arguments and evidence to prove or substantiate their position. But neither side appears to have won the contest or is even showing signs of making progress on it. This problem is not unique, especially in the field of philosophy. We have seen it before. In fact, it is similar to the problem that Kant had addressed and resolved in his *Critique of Pure Reason*. Kant's first critique[7] famously sought to address what had been a deadlock in modern European philosophy—the seemingly irresolvable debate between the Rationalists and Empiricists regarding the origins of human knowledge. And in response to this dispute, Kant changed everything by simply altering the terms of the debate: "Hitherto it has been assumed that all our knowledge must conform to objects. But all attempts to extend our knowledge of objects by establishing something in regard to them *a priori*, by means of concepts, have, on this assumption, ended in failure. We must therefore make trial whether we may not have more success in the task of metaphysics, if we suppose that objects must conform to our knowledge" (Kant 1965, Bxvi; emphasis in original).

When debate seems to get stalled in an irresolvable stalemate or eternal recurrence of the same, throwing more argumentative effort at the dispute only perpetuates the existing problem. A better solution may be to alter the terms of the debate itself. For Kant, this meant not asking how knowledge conforms to objects but rather how objects correspond to our modes of knowing. And that critical pivot has made all the difference. So like Kant, we can investigate whether we may not have more success in the task of responding to the opportunities and challenges of robots and AI systems if we shift our focus and the mode of inquiry. Instead of trying to resolve the problem as it is currently formulated—that is, deciding whether robots, AI applications, and other kinds of artifacts are things or persons—the deconstruction of this way of thinking takes the very conceptual opposition that had distinguished person from thing as the problem.

3. Ethnocentrism. Finally, the seemingly natural opposition between person and thing, which "for so long compressed and continues to compress human experience into the confines of this exclusionary binary equation" (Esposito 2015, 4), proceeds from a distinctly Western way of thinking, "genetically composed of the confluence between Greek philosophy, Roman law, and the Christian conception" (3). This way of organizing things—this method of dividing up all of existence into one of two mutually exclusive types—is not only exported around the world through colonial conquest and religious conversion but has, for better or worse, dominated the entire domain.

The person/thing dichotomy, however, is not some universal truth. It is a specific way of organizing things that is rooted in and determinative of Western thought.[8] It is, therefore, not some Platonic form that would be universally true for all time. In fact, other cultures and traditions approach these challenges in ways that are significantly different and organized otherwise. Consider, for example, what Suzanne Kite explains by way of Lakota ontologies in the collaborative written essay "Making Kin with the Machines":

> How can humanity create relations with AI without an ontology that defines who can be our relations? Humans are surrounded by objects that are not understood to be intelligent or even alive, and seen as unworthy of relationships. In order to create relations with any non-human entity, not just entities which are human-like, the first steps are to acknowledge, understand, and know that non-humans are beings in the first place. Lakota ontologies already include forms of being which are outside of humanity. Lakota cosmologies provide the context to generate an ethics relating humans to the world and everything in it. These ways of knowing are essential tools for humanity to create relations with the non-human and they are deeply contextual. As such, communication through and between objects requires a contextualist ethics which acknowledges the ontological status of all beings. (Lewis et al. 2018)

Critically questioning and challenging the person/thing dichotomy is not just a way of recognizing and coming to terms with ethnocentrism, colonialism, and the multifaceted legacy of what Kim Tallbear calls "whitestream disciplinary thinking and ontologies" (Muñoz et al. 2015, 230). It is also a way to begin opening the debate about AI systems and robots to other perspectives, traditions, and modes of thinking. These alternatives provide other ways of responding to and taking responsibility for what José Esteban Muñoz calls "the active self-attunement to life as varied and unsorted correspondences, collisions, intermeshings, and accords between people and nonhuman objects, things, formations, and clusterings" (Muñoz et al. 2015, 210). For this reason, it should be no surprise that deconstruction has not found a home (or found itself *at* home) in the established institutions and departments of philosophy but has instead enjoyed a more hospitable reception in other areas of endeavor: literary criticism, feminism, postcolonialism, posthumanism, queer theory, and so on.

1.6 Preview/Overview

We will begin with *things*, which is the rather unimposing title to the second chapter. Beginning here seems almost unnecessary and obvious. We all kind of know what things are, such that asking a question like "What is a thing?" seems to be impertinent and immaterial. But as the German philosopher Martin Heidegger pointed out, this is precisely the problem. Because things are already familiar—perhaps too familiar—we have little or no critical distance on them as things. This is because, as Heidegger (1962) explains, things are not typically disclosed to us as things but encountered as objects. In other words, things are not experienced as mere entities laying around out there in the world. They are always pragmatically situated and characterized in terms of our involvements and interactions with the world in which we live. For this reason, things are first and foremost made available to us or revealed as objects for a subject. They are objectified. The chapter, therefore, examines how the two sides of the debate—the Critics and the Advocates—mobilize this objective understanding of things to construct and develop their arguments. One side, not surprisingly, argues that robots and AI systems are objects or instruments for us to use for our own purposes and objectives. The other side asserts that there is something about these things that makes them more or at least interrupts this objectivist and instrumentalist way of thinking. Although each side offers persuasive arguments and extensive evidence to support their claims, neither is able to gain the upper hand. And this leaves us with uncertainty regarding the reification of robots, AI systems, and other artifacts.

The third chapter, then, takes up the other alternative and inquires whether these things might not be better and more successfully understood as persons. Again, the question "Are robots persons?" seems rather direct and immediately understandable. But as was the case with *thing*, the concept of *person* also turns out to be far more interesting and complicated than one might anticipate. Typically, when we use the word *person*, we are referring to another human being. But this seemingly natural and everyday understanding is not entirely accurate. The word and concept *person* has a long and rather complicated history in both philosophy and law. It is originally derived from the Latin *persona*, which referred to the mask worn by actors in a stage play. In its original form, what we call a *person* would have been closer to what is meant by a word like *character* or *role*. As a result, *person* does not just denote an individual human being but also, more accurately, names the role that one plays or is assigned within the context of a social situation or performance. This can be seen especially in law, where who is considered a legal person has not been limited to human individuals but can also be extended to other kinds of nonhuman entities: corporations, organizations, ships, animals, and the natural environment. How this happens, what it means, and how it organizes the terms of the debate is the subject of chapter three.

The fourth and fifth chapters critically investigate the debate regarding personification of robots, AI systems, and other artifacts. In other words, once we know what *person* designates and means, we can then examine how the two sides in the debate have argued for and against extending the title of person to these various nonhuman entities. Chapter four takes up the subject of *natural person*, which typically denotes human individuals or other entities who can be considered persons by nature. Whether something is or is not a natural person is usually decided on the basis of individual properties and capabilities—that is, rationality, consciousness, sentience, and the like. So there is a kind of litmus test for achieving recognition as a person. We (and who is implicated in or interpellated[9] by this first-person-plural pronoun will not be insignificant) first define the criteria for what makes an entity a person. In other words, we devise a standard by deciding what we believe are the necessary person-making properties or capabilities. We then use this standard to test and evaluate whether some entity is a natural person or not. So the question of whether robots or AI systems could ever become natural persons is one that we get to evaluate based on criteria that we get to define.

The Critics respond to this question by trying to limit who can be included in the category of person. According to these arguments, robots, AI systems, and other artifacts are just technological devices and machines.

No matter how sophisticated they may become or seem to be, they will never achieve any of the person-making properties, like consciousness, rationality, or sentience. And because of this, robots and other kinds of artifacts will always fail to meet the necessary criteria and can therefore be objectively and justifiably refused the title of person. The other side in the debate takes an entirely different and polar opposite position, arguing that artifacts like AI systems, robots, and autonomous technologies either are or will be able to achieve the benchmarks of personhood. If and when (and in the hands of these Advocates, it is more often than not a matter of when) this happens, then withholding the title of person from these things would be unethical and unjust.

Chapter five does a similar kind of analysis for the concept of a legal person. Unlike a natural person, which is grounded in the ontological conditions or essential properties of the individual entity, a legal person is a socially constructed and conferred recognition. In other words, something becomes someone not because of their essential nature but because they are recognized by others as having a particular status. To be a person, then, means that one is recognized as a subject under the law, possessing both responsibilities and rights within a particular legal construct or institution. If the paradigmatic natural person is the human being in possession of a set of natural capabilities that make one a person, then the paradigmatic legal person is the corporation, which is a person not by its nature but because it is recognized and situated within the law as a subject of the law.

The question confronted in the face or the faceplate of the robot or other seemingly intelligent and/or socially interactive artifact is whether it would make sense to extend the category of legal person to these other kinds of entities. Responses to this question once again divide into two seemingly opposed and irreconcilable positions. Those who are critical of this proposal argue that extending the recognition of person to these technological artifacts, although clearly possible and entirely legal, is wrong and should not be allowed to happen. And it is wrong because of the negative effects and potentially dangerous consequences this decision would have on and for us and our legal systems. Those situated on the other side of the debate take an entirely different position on this question, arguing that extending legal personhood or personality to robots and other intelligent (or at least seemingly intelligent) artifacts will be necessary for integrating these technologies into our moral and legal systems.

By the time we get to the end of the fifth chapter, it will become evident that the debate is ultimately irresolvable. Both sides present good arguments and provide substantial evidence to support their positions. Unfortunately, this produces a version of Hegel's "bad infinity," with very little progress being made on resolving the question at hand. It is in response to this difficulty that the sixth chapter identifies and critically evaluates viable alternatives.

In the face of these seemingly irreducible either/or dilemmas, one solution—and a rather popular one at that—is to synthesize a third alternative that is either neither/nor or both/and. This is standard operating procedure in all kinds of disputes and debates. In politics, for instance, the tension between the right and the left is typically mediated by a third alternative that has been called the *center* or *radical middle*. In philosophy, the difference that separates binary opposites, like being and nothing, comes to be resolved in and by a third term, *becoming*, which *sublates* (Hegel's watchword, meaning "to overcome and preserve") the difference between the one and the other. And in many legal systems, the opposition between the mutually exclusive gender categories of male and female has been successfully challenged and resolved by formulating a third option: nonbinary.

A similar strategy has been proposed for resolving the person/thing debate with robots and AI, and that solution goes by the name *slavery*. Already in Roman times, slaves were regarded as something more than a mere thing but not quite a full person. They occupied a position that was situated in between the one and the other, being both thing and person. And there has been, in both the legal and philosophical literature, a surprising number of serious proposals arguing for instituting what can only be called Slavery 2.0. Repurposing existing slave law to respond to the moral and legal challenges of robots might seem like a workable solution, but the difficult history of human slavery and its horrific social and political consequences actually produce more problems than they can possibly resolve. The sixth chapter provides a thorough critique of these "robots should be slaves" proposals, demonstrating how this supposed solution to the person/thing dichotomy actually produces bigger problems—that is, more and significantly worse problems than it can possibly begin to resolve.

The problem, then, is not deciding whether robots and other kinds of artifacts are persons, things, or some third alternative that tries to split the difference. The real problem is the binary logic that differentiates person from thing in the first place. This fundamental distinction is not some natural condition of things: it is an artificially constructed dichotomy that is context specific and only subsequently universalized by the exercise of socio-intellectual-political power. The seventh and final chapter deconstructs this way of thinking. It does so by way of a double gesture that involves (1) inversion of the person/thing dichotomy and (2) emergence of a nondialectical third term that exceeds the grasp of this entire conceptual order. The book ends, therefore, by developing the terms and conditions of this deconstructive alternative in an effort to respond to the alterity that is manifest in the face of the robot. But this alternative, because it deconstructs the ruling conceptual order—that is, the person/thing dichotomy that has organized both law and ethics for close to two millennia—will have repercussions that reverberate both forward and backward through time, requiring a thorough reconfiguration of moral and legal ontology.

1.7 Final Words

Often it is the smallest of things that matters most, like the difference between two seemingly insignificant words—*who* and *what*. But everything, as Derrida (2005, 80) insightfully points out, depends on this difference. In the face of others—not just other human beings but animals, the things of the natural environment, and artifacts—we are called upon to make and are responsible for a crucial decision. We decide (or perhaps better, have granted to ourselves the power to decide) between *who* counts as another socially significant subject with rights and responsibilities that need to be respected and *what* remains a mere thing that can be used and even abused as we see fit. It's a seemingly small difference, but it makes all the difference.

It has, for instance, justified some members of the human species asserting their presumed authority over other members of same species (and typically on the basis of arbitrary and highly prejudicial criteria like gender, race, ethnicity, religion, geographical location, or cultural traditions), turning them into instruments and things to be bought and sold as property and even disposed of with little or no hesitation. It has allowed human communities to excuse horrific mistreatment of other life forms on the basis that they are not persons and therefore can only be things that serve our interests, needs, and desires. And it has led us—not just human beings but all living and nonliving things on planet Earth—into a climate crisis that constitutes a fundamental existential challenge, as natural human persons and organizational and corporate legal persons have presumably decided that other things are nothing more than "natural resources" and "raw materials" that can be used, consumed, and exploited.

The true potential of the moral and legal challenges that are confronted in the face or the faceplate of robots and other artifacts is that they destabilize this ontological order that has permitted us (and again, who is included in and excluded from this first-person-plural pronoun is not immaterial) to divide all of existence into the mutually exclusive categories of person or thing. It is, therefore, the robot that might deliver us from ourselves, demanding that we begin to take seriously our responses to and responsibilities for other Things—Things that exist and have always existed outside and beyond the limitations of the arbitrary division that differentiates who is person from what is a thing.

2 Things

One thing that both sides in the debate seem to agree on is the state and status of things. Things are categorically different from persons. Whereas the category of person refers to those others who count and have moral and legal status, things do not. They are objects as opposed to subjects and therefore can be used, abused, and disposed of as we decide and see fit. This fundamental distinction not only sounds right, it's the law. As Esposito (2015, 1–2) remarks: "When the Roman jurist Gaius, in his *Institutes*, identified persons and things as the two categories that along with actions constitute the subject matter of the law, he did nothing more than give legal value to a criterion that was already widely accepted." Because of this, the two sides in the debate do not dispute what things are. Where they diverge and what they argue about is whether robots and other seemingly intelligent and/or socially situated artifacts are things, like any of the other things with which we are already familiar, or something that is (or can become) more than what is typically understood and designated by the term *thing*.

2.1 What Is a Thing?

There is, it seems, nothing particularly interesting or extraordinary about things. We all know what things are; we deal with them every day. But as the twentieth-century German philosopher Martin Heidegger (1962) pointed out, this immediacy and proximity is precisely the problem. Because things are so ubiquitous, common, and useful, we typically cannot see them as things. Marshall McLuhan and Quentin Fiore (2001, 175) cleverly explained the problem this way: "one thing about which fish know exactly nothing is water." Like fish that cannot perceive the water in which they live and operate, we are often unable to see the things that are closest to us and comprise the very milieu of our everyday existence.

In response to this difficulty, Heidegger expends substantial effort investigating what things are and why things seem to be more difficult than they initially appear. In fact, the question "What is a thing?" is one of the principal concerns and an organizing principle of Heidegger's phenomenological project. And this attention to things begins right at the beginning of his 1927 magnum opus, *Being and Time*: "The Greeks had an appropriate term for 'Things': πράγματα [*pragmata*]—that is to say, that which one has to do with in one's concernful dealings (πραξις). But ontologically, the specific 'pragmatic' character of the πράγματα is just what the Greeks left in obscurity; they thought of these 'proximally' as 'mere Things.' We shall call those entities which we encounter in concern '*equipment*'" (Heidegger 1962, 96–97).

According to Heidegger's analysis, things are not, at least not initially, experienced as mere entities out there in the world. Things are not just there. They are always pragmatically situated and characterized in terms of our practical involvements and interactions with the world in which we live. For this reason, things are first and foremost revealed as pieces of "equipment" or "instruments" that are useful for our endeavors and objectives. "The ontological status or the kind of being that belongs to such equipment," Heidegger (1962, 98) explains, "is primarily exhibited as 'ready-to-hand' or *Zuhandenheit*, meaning that something becomes what it is or acquires its properly 'thingly character' when we use it for some particular purpose."

Explanations like this are what makes reading Heidegger both interesting and frustratingly difficult. So let's break it down. According to Heidegger, the fundamental ontological status, or mode of being, that belongs to things is primarily exhibited as "ready-to-hand," meaning that something becomes what it is or acquires its properly "thingly character" in coming to be put to use for some particular purpose or end—a purpose that we define and operationalize. A hammer, one of Heidegger's principal examples, is for building a house to provide shelter from the elements; a pen is for writing a book like *Being and Time*; a shoe is designed to support the activity of walking. Everything is what it is in having a "for which" or a destination to which it is always and already referred. Everything therefore is primarily revealed as being a tool or an object—that is, a means to an end—which is useful for our purposes, needs, and projects.

The term *object* is crucial in this context. Things are always and already objectified by being situated in opposition to and placed in the service of a human subject.[1] This mode of existence—what Graham Harman (2002) calls *tool-being*—applies not just to artifacts, like hammers, pens, and shoes. It also describes the basic existential condition of natural objects, which are, as Heidegger (1962, 100) explains, discovered in the process of being put to use: "The wood is a forest of timber, the mountain a quarry of rock, the river is water-power, the wind is wind 'in the sails.'" Everything therefore exists and becomes what it is insofar as it is useful for some human-defined purpose. Things are not just out there in a kind of raw and naked state but come to be what they are in terms of how they are objectified and put to work or used by us as equipment for living.

This is precisely what makes things so difficult to perceive. Whatever is ready-to-hand is essentially transparent, unremarkable, and even invisible. "The peculiarity," Heidegger (1962, 99) writes, "of what is proximally ready-to-hand is that, in its readiness-to-hand, it must as it were, withdraw in order to be ready-to-hand quite authentically. That with which our everyday dealings proximally dwell is not the tools themselves. On the contrary, that with which we concern ourselves primarily is the work." In being objectified, we do not notice the thing as such; we see through it to the purpose that it is intended to serve.[2] Or as Michael Zimmerman (1990, 139) explains by way of Heidegger's hammer, "In hammering away at the sole of a shoe, the cobbler *does not notice the hammer*. Instead, the tool is in effect transparent as an extension of his hand. . . . For tools to work right, they must be 'invisible,' in the sense that they disappear in favor of the work being done."

This understanding of things can be correlated with what philosophers of technology call the "instrumental theory of technology," which Heidegger subsequently addressed in *The Question Concerning Technology*. "We ask the question concerning technology," Heidegger (1977, 4–5) writes, "when we ask what it is. Everyone knows the two statements that answer our question. One says: Technology is a means to an end. The other says: Technology is a human activity. The two definitions of technology belong together. For to posit ends and procure and utilize the means to them is a human activity." According to this explanation, the presumed role and function of any kind of technological object, whether it be a simple mechanical device, like a corkscrew or bicycle, or a digital computer or smartphone, is that it is a means employed by human users for specific ends. Heidegger terms this particular characterization of technology the *instrumental definition* and indicates that it forms what is considered to be the correct understanding of any kind of technological contrivance.

As Andrew Feenberg (1991, 5) has neatly summarized it, "The instrumentalist theory offers the most widely accepted view of technology. It is based on the common sense idea that technologies are 'tools' standing ready to serve the purposes of users." And because a tool or an instrument "is deemed 'neutral,' without valuative content of its own," then a technological thing is evaluated not in and of itself but on the basis of the particular employments that have been operationalized by its human designer, manufacturer, or user. For this reason, technology, as Jean-François Lyotard (1984, 44) concludes, "is a game pertaining not to the true, the just, or the beautiful, etc., but to efficiency: a technical 'move' is 'good' when it does better and/or expends less energy than another."

This objectivist or instrumentalist way of thinking also finds expression in the way the legal literature deals with things. "By thing (*res*)," German jurist Anton Thibaut (1855, 16; emphasis in original) explains, "is meant whatever neither is nor can be the *subject* of a legal relation, but yet may be the *object* of a legal transaction and so mediately the object of a right." Formulated in this way, things are not considered the subject of legal relationships involving rights and obligations. They are, however, the objects of these relationships and have instrumental value due to the way they mediate between persons who are the direct and proper subjects of rights and obligations. Consequently, things, like robots, AI, and other technological artifacts, no matter how sophisticated or autonomous they appear or are designed to be, are not and cannot be the subject of moral or legal concern. They are *just* tools or instruments, objects to be used by subjects—namely, human persons.

Finally, and perhaps most importantly for the analysis that follows, this characterization of things, though seemingly correct and virtually beyond question, is not some natural condition or universal truth that is valid for all times and places. It is the product (i.e., an artifact) of a particular cultural tradition that comes to be elevated to the standpoint of an assumed universal truth through the exercise of intellectual and sociopolitical power. As Esposito (2015, 3) is careful to note and point out, this way of thinking about things is the result of a confluence of "Greek philosophy, Roman law, and the Christian conception." And it is, as Michael Szollosy (2017, 156) recognizes, in the face or the faceplate of the robot that we can begin to perceive the significance and consequences of this ethnocentrism: "Robots are considered to be machines, and therefore merely objects. In the European Christian tradition, such non-living, or even non-human objects, are considered lesser beings on the basis that they do not have a soul; that intangible, metaphysical property unique to life or, in most articulations, unique specifically to humans. . . . Though one could argue that Europe is no longer beholden to Christianity, Europe's (and America's) Christian values are constantly on display, and this assumption is obvious even in contemporary, completely secular European legal and ethical frameworks."

Although the vast majority of existing publications on the subject of robots, AI systems, and other artifacts adhere to and proceed according to the standard instrumentalist way of thinking about things, we (and in this case, the first-person-plural pronoun incorporates anyone and everyone who already occupies and operates from an assumed and largely unquestioned Western philosophical position) will need to be vigilant and continually remind ourselves that this way of proceeding is itself the product of a specific tradition and its hegemony. Keeping this in mind not only will provide the occasion and raison d'être for deconstructing things, it will also help us account for the fact that robots—real robots, right here and right now—have already begun to challenge this specific and rather limited way of thinking about things.

2.2   The Thing with Robots

Even though things are initially disclosed and revealed to us in the mode of being Heidegger calls *Zuhandenheit* (i.e., instruments that are useful or handy for our purposes and endeavors), things do not necessarily end here. They can also, as Heidegger (1962, 103) explains, be subsequently present-at-hand, or *Vorhandenheit*, revealing themselves to us as things that are or become, for one reason or another, *un-ready-to-hand* or obtrusive. This is evidently what happens with many robotic things, like chatbots, machine learning algorithms, and social robots, insofar as they appear to interrupt or otherwise impede the smooth functioning of the instrumentalist way of thinking. The thing with robots, AI systems, and similar artifacts is that they are or often appear to be something other than instruments. Consider three examples.

First, there are things that talk and appear to use language in ways that are intelligible, like chatbots, digital assistants, and large language models. In fact, the emulation of human conversational behaviors has been the defining condition of machine intelligence since Alan Turing's agenda-setting paper from 1950, "Computing Machinery and Intelligence." Users of these systems often find it difficult to distinguish between human communication mediated through the instrumentality of a computer and conversational behaviors generated by the machine itself—so much so that there have been a number of legislative efforts, stipulating that any and all implementations of what Simone Natale (2021, 11) calls "communicative AI" disclose their true artificial nature prior to interacting with human users, a kind of "Warning: I am AI" label. Unlike artificial general intelligence (AGI), which would presumably occupy a subject position reasonably close to that of another human person, these actually existing and ostensibly mindless but very loquacious things simply muddy the water (which is probably worse) by complicating and leaving undecided questions regarding who or what is actually doing the talking.

Second, there are things that are deliberately designed to exceed (or at least seem to exceed) the comprehension, oversight, and control of their human developers—things like DeepMind's AlphaGo and its successor AlphaZero, self-driving automobiles and lethal autonomous weapons systems (LAWS), and many other machine learning implementations and algorithms. Although the extent to which one might reasonably assign "autonomy" to these mechanisms remains a contested issue, what is not debated is that the rules of the game seem to be in flux. As Andreas Matthias (2004, 177) points out, summarizing his survey of learning automata: "Presently there are machines in development or already in use which are able to decide on a course of action and to act without human intervention. The rules by which they act are not fixed during the production process, but can be changed during the operation of the machine, by the machine itself. This is what we call machine learning." The instrumental theory, which effectively tethers machine decision and action to human persons, no longer adequately applies to or fully explains mechanisms that have been deliberately designed to operate and exhibit some form, no matter how rudimentary, of independent action or autonomous decision making. Contrary to the usual instrumentalist way of thinking, we now have things that have (or at least seem to possess) a mind of their own.

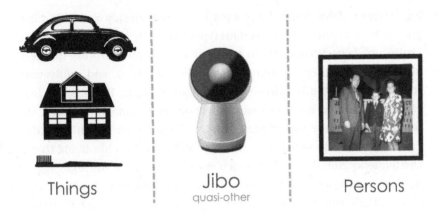

Figure 2.1
The social robot is located in a position that is in between the two categories of *thing* and *person*. Original image by the author.

And third, there are things that are designed to be and promoted as being something that is more than a mere thing, like robot companions and social robots. In 2014, for instance, social robotics pioneer Cynthia Breazeal introduced the Jibo robot with the following explanation: "This is your car. This is your house. This is your toothbrush. These are your things. But these [and the camera zooms into a family photograph] are the things that matter. And somewhere in between is this guy. Introducing Jibo, the world's first family robot" (Jibo 2014). According to this ontological classification, Jibo is not just another instrument or tool, like the automobile or toothbrush. But he/she/it (and the choice of pronoun is not unimportant) is also not quite another member of the family pictured in the photograph. The social robot inhabits a place in between the two categories of thing and person (figure 2.1). It is a kind of "quasi-other" (Ihde 1990, 107). This is, it should be noted, not necessarily unprecedented. We are already familiar with other entities that occupy a similar ambivalent social position, like the family dog.

In fact, as Kate Darling (2021) has effectively argued, animals and pets in particular provide a good template for understanding the changing nature of things. "Looking at state of the art technology," Darling (2012, 1) writes in one of her early essays on the subject, "our robots are nowhere close to the intelligence and complexity of humans or animals, nor will they reach this stage in the near future. And yet, while it seems far-fetched for a robot's legal status to differ from that of a toaster, there is already a notable difference in how we interact with certain types of robotic objects." This occurs, Darling explains, because of our tendency to anthropomorphize things by projecting onto them cognitive capabilities, emotions, and motivations that do not necessarily exist in the mechanism per se. But it is this inescapable emotional reaction that necessitates new forms of response and responsibility in the face of these things. "It's important," Darling (2021, 154) concludes, "to understand that we *will* treat robots like living things. Our tendency runs deep, and as much as we could decry it and argue against it, it's not going away. When it comes to interacting with machines, even more so than with animals, we *know* that we're projecting something that's not there, and we do it anyway."

All these "things" (though that word may not be entirely suitable in this context, hence the quotes) seem to resist efforts at reification and/or impede classification as instruments and objects. They do so because they have been designed to be or to appear to be (and the difference between "being" and "appearing to be" will not be insignificant) intelligent, autonomous, or social. Responses to this challenge (or opportunity—it all depends on how you approach and interpret it) pull in two apparently different and opposite directions.

2.2.1 The Critics and Their Arguments

In the face of these things, it is possible to respond as we typically have, treating these increasingly communicative, autonomous, and seemingly social mechanisms as mere instruments or objects. "Computer systems," as Deborah Johnson (2006, 197) has argued, "are produced, distributed, and used by people engaged in social practices and meaningful pursuits. This is as true of current computer systems as it will be of future computer systems. No matter how independently, automatic, and interactive computer systems of the future behave, they will be the products (direct or indirect) of human behavior, human social institutions, and human decision." This mode of argument is persuasive precisely because it draws on and is underwritten by the usual understanding of things. Things—no matter how sophisticated, intelligent, and social they are, appear to be, or may become—are and will continue to be tools of human decision and action, nothing more.

If something goes wrong (or goes right) because of the actions or inactions of a robot, AI, or some other seemingly intelligent thing, there is always someone who is ultimately responsible for what happens with it. Finding that person (or persons) may require sorting through layer upon layer of technological mediation, but there is always someone—specifically, some human someone—who is presumed to be responsible and accountable for it. According to this way of thinking, all artifacts, no matter how sophisticated or interactive they appear to be, are actually *Wizard of Oz* technology.[3] There is always a "man behind the curtain," pulling the strings and responsible for what happens. And this line of reasoning is entirely consistent with current legal practices. "As a tool for use by human beings," Matthew Gladden (2016, 184) argues, "questions of legal responsibility . . . revolve around well-established questions of product liability for design defects (Calverley 2008, 533; Datteri 2013) on the part of its producer, professional malpractice on the part of its human operator, and, at a more generalized level, political responsibility for those legislative and licensing bodies that allowed such devices to be created and used."

The same kind of thinking and mode of argument applies to the question concerning moral or legal status. As Johannes Marx and Christine Tiefensee (2015, 83) explain in the essay "Of Animals, Robots and Men": "Robots are nothing more than machines, or tools, that were designed to fulfill a specific function. These machines have no interests or desires; they do not make choices or pursue life plans; they do not interpret, interact with and learn about the world. Rather than engaging in autonomous decision-making on the basis of self-developed objectives and interpretations of their surroundings, all they do is execute a preinstalled programme. In short, robots are inanimate automatons, not autonomous agents. As such, they are not even the kind of object which could have a moral status." This is a near-perfect restatement of the instrumentalist argument. Robots and other things like this are nothing more than tools and instruments that are destined for and designed to fulfill our needs and projects. They do not have their own interests or desires. They are not really autonomous agents. They are just preprogrammed objects that are designed to do what we decide and direct them to do. Consequently, robots, AI systems, and other kinds of seemingly intelligent artifacts are simply not the right kind of thing to have rights or obligations. A phrase like *robot rights* would, on this account, be a contradiction in terms or an oxymoron.

Arguments like this appear throughout the literature. Andrea Bertolini (2013, 235), for instance, has maintained that robots—even highly sophisticated and seemingly autonomous devices—should be "deemed objects—more precisely, artefacts created by human design and labour, for the purpose of serving identifiable human needs." This goes not only for industrial and service robots but also, and perhaps more importantly, for social or companion robots in domestic settings: "Robot companions, just like all existing robots, are to be deemed objects, more precisely, artefacts created by human design and labor, for the purpose of serving identifiable human needs. In brief, they are, to all—legal and ethical—effects and purposes, to be deemed products" (Bertolini and Aiello 2018, 130). The reasoning behind and supporting this declaration is seemingly simple: "So long as robots do not achieve self-consciousness," Bertolini (2013, 235) concludes, "they cannot be deemed moral agents or autonomous—in a strong sense—beings. Short of that capacity there is no logical, moral or philosophical—and thus not even legal—necessity to consider them subjects of law and bestow individual rights on them."

Similar assertions have been issued in the philosophical literature on the subject. In the essay "On the Moral Status of Social Robots," for example, Kęstutis Mosakas (2021b, 429) deploys what he calls the *consciousness criterion* to argue that robots, and social robots in particular, "should not be regarded as proper objects of moral concern unless and until they become capable of having conscious experience." Until that threshold condition is achieved, he (like Bertolini) concludes that these things are and will remain mere things, nothing more.

John Basl and Joseph Bowen (2020, 296), in their entry for the *Oxford Handbook of Ethics of AI*, elaborate on this insight, providing the following argument:

1. AI is a potential rights holder only if it is a bearer of well-being.
2. AI is a bearer of well-being only if it is conscious.
3. AI is not conscious.
4. So, AI cannot be said to have rights.

According to this argumentative procedure, something is a subject of moral concern—a potential holder of both rights and obligations and not just an object—if and only if it fulfills what Mosakus calls the consciousness criterion. But robots and AI systems—no matter how sophisticated their design and seemingly intelligent their behaviors—are just instruments and objects that, by definition, lack any kind of consciousness whatsoever. For this reason, these artifacts cannot, by definition, have rights and should not be said to possess or be in need of possessing rights.

A similar version of this line of reasoning has been developed and presented by Abeba Birhane and Jelle van Dijk (2020) in an essay published in *Noema*, a more popular online publication aimed at a general audience: "We should not be seeing robots and AI algorithms as replicas of ourselves, nor as entities that can be granted or denied rights. Instead, we should see them as they actually exist in the world: as an increasingly influential element within the socio-material context that human beings have produced as part of the process of making sense of the world. To put it another way, if AI algorithms are first and foremost things we use to think with and not, in themselves, thinking things, then arguing for artificially intelligent agents like robots to have rights becomes problematic." Although not explicitly identified as such, this is the standard instrumentalist argument. AI systems and robots are things—influential and important things that we use to make sense of and think about our world. Nevertheless, these things, as mere things, are essentially and categorically different from us. We are, by contrast, not just "thinking things," which had been Descartes's formulation, but living bodies who "laugh, bite, eat, gesture, breathe, give birth and feel pain, anger, frustration and happiness" (Birhane and van Dijke 2020). As a result of this undeniable and irreducible ontological difference, not only can robots, AI systems, and other artifacts not be the subjects of rights but the question whether they can be granted or denied such status is itself a problem. Robots and AI algorithms are categorically the wrong kinds of things for rights. Thus, the question of rights and their protections simply does not apply.

Vincent C. Müller develops another iteration of this in his journal article titled "Is It Time for Robot Rights? Moral Status in Artificial Entities." In response to the question in the title of the essay, Müller (2021, 579) answers in the negative: "There is no reason to think that robot rights are an issue now." In other words, at this particular point in time, given the currently available technology at our disposal, robots and AI systems are simply not the right kinds of things for rights. This conclusion, it is important to note, is tempered by two concessions. The first is temporal or time-based. "We do not claim," Müller (584) writes, "to have made a general case against moral status of robots, quite the contrary: We just tried to work out the assumptions in the proposals for attributing moral status to them right now—and find these problematic. One should attribute moral status to robots when they fulfill the criteria."

The second is epistemological and concerns the characterization of the criteria. "It is good to keep in mind that the criteria for moral status admit to a degree of vagueness, so we should not expect the set of 'objects with moral status' to have a sharp border (e.g. very young humans are not moral agents, but they can gradually grow to be agents). But even without sharp borders, there are objects that clearly fall into that set, and objects that do not" (Müller 2021, 584). Despite these two important clarifications, the argument still proceeds on the standard assumption that ontology (i.e., what something is) determines its moral status (i.e., how it is to be treated). For now, at least, robots are not the kinds of things that can achieve what is necessary to have moral status (however that actually comes to be defined) and to be subjects of rights or obligations. They are, for the time being, mere things.

All of these efforts operationalize and advance a similar kind of argument: Technological artifacts, like robots and AI, no matter how sophisticated or autonomous they appear or are designed to be, are not and cannot be moral or legal subjects. They are (at least as they exist right here and right now) *just* tools or instruments to be used by moral and legal subjects—namely, human persons. This means not only that robots, AI systems, and other kinds of artifacts cannot have rights but also that these things are not the right kinds of things to ever have (or not have) rights. As Birhane and Van Dijk (2020, 1) have succinctly described: "We argue not just to deny robots 'rights,' but to deny that robots, as artifacts emerging out of and mediating human being, are the kinds of things that could be granted rights in the first place." For this reason, the mere idea of robot rights or the question concerning the moral and legal status of artifacts is determined to be a mistake and a category error. Ultimately, it's pure nonsense, and to entertain the idea is a waste of time and effort that risks distracting us from more important things.

2.2.2 The Advocates and Their Arguments

Those situated on the other side of the debate do not dispute the instrumental theory. They simply argue that robots, AI systems, and other things like them do not necessarily fit within its theoretical framework and conform to its provisions. Robots, AI systems, and other artifacts, it is argued, are not like other things. There is a crucial difference. These things are capable—either right now in their current form or in the not-too-distant future—of possessing (or at least giving seemingly convincing evidence of possessing) the right set of psychological properties and social behaviors that would qualify them to be not just objects but other kinds of moral and legal subjects, who can and should have access to both rights and obligations. Even though they might not be anything like a human being, they would be, to use terminology developed by German Catholic philosopher Robert Spaemann (2006), not *something* but *someone*, who has a legitimate claim to rights and their protections. Thus, continuing to categorize and treat them as mere things or objects would be wrong.

What is perhaps not surprising is the fact that the number of publications coalescing on this side of the debate easily exceed and outnumber those situated on the other side.[4] And there is a good reason for this. The underlying ontological distinction that divides person from thing—along with the instrumental theory of technology that further explicates what things are and are not—is already taken to be self-evident, obvious, and intuitively correct. For this reason, robots and AI systems, which are things designed, built, and used by human persons, do not have nor need rights. This way of thinking—persons are subjects of rights and obligations and things are objects that are not—already sounds correct and beyond question. It is status quo, the orthodox account, the standard operating procedure.

Consequently, arguments in support of this position—that is, efforts to prove what is already taken to be correct and unassailable—seem to be not just unnecessary but little more than profound statements of the obvious. They only become necessary in the face of and in response to opposing positions that promote or advance the seemingly counterintuitive conclusion that things like robots and AI systems can and/or should have rights, obligations, or both. So if there appears to be less argumentative effort assembled on the side of the Critics, it is because their way of thinking about things is already assumed to be right. In more formalistic/legal language, one could say that the burden of proof resides on the side of the Advocates, because they seek to promote a position that is both counterintuitive and significantly different from the usual, accepted ways of thinking.

These efforts to question and challenge the status quo typically take the form of a conditional statement, or what Jesse De Pagter (2021, 6) calls *if-and-then rhetorics*. "Today," Christian Neuhäuser (2015, 133) explains, "many people believe that all sentient beings have moral claims. But because people are not only sentient, but also reasonable, they have a higher moral status as compared to other animals. At least this is the prevailing opinion within the discussion. According to this position, humans do not only have moral claims of only relative importance but also inviolable moral rights because they possess dignity. If robots become sentient one day, they will probably have to be granted moral claims." The operative words here are *if* and *probably*. *If* robots achieve a certain level of cognitive capability—that is, if they possess some morally relevant capacity like consciousness or sentience (or whatever else is decided to be the qualifying criteria)—then they *probably* will have a legitimate claim to moral status and should have rights. Conditional statements like this are usually future oriented, and there are several different iterations of this kind of argument available in the literature.

Animal rights innovator Peter Singer has weighed in on this question a number of times. In an article written with Agata Sagan, titled "Robot Rights?" and published in *Project Syndicate* (2009a) and the *Guardian* (2009b) Singer offered the following assessment: "If the robot was designed to have human-like capacities that might incidentally give rise to consciousness, we would have a good reason to think that it really was conscious. At that point, the movement for robot rights would begin." For Singer and Sagan, the cause of robot rights—the point at which it would no longer be justified to treat these artifacts as mere instruments and objects—is dependent upon a prior determination of consciousness. Once that threshold condition is met, robot rights is not just possible but highly probable.

In a video interview with Big Think (2019), Singer elaborates his thinking on this subject, connecting the dots between his own agenda-setting research in animal rights to the nascent efforts in and discussions about the moral condition and status of artifacts:

> I've argued that throughout history we have expanded the circle of moral concern from initially it just being our own tribe to a nation, race, and now all human beings. And I've been arguing for expanding beyond just human beings to all sentient creatures, all beings capable of feeling pain, enjoying their life, feeling miserable. And that obviously includes many nonhuman animals. If we get to create robots that are also capable of feeling pain then that will be somewhere else that we have to push the circle of moral concern backwards because I certainly think we would have to include them in our moral concern once we've

actually created beings with capacities, desires, wants, enjoyments, miseries that are similar to ours.

Singer's argument, here and in the article written with Sagan, is clearly anchored in what Mark Coeckelbergh (2012, 13) calls the *properties approach* to deciding questions of moral status and rights.[5] And Singer is, if anything, consistent in his application of this way of proceeding. If we create robots that have the ability to feel pain (again stated as a condition), then we will need to include them in the community of moral subjects, because things that feel pain are not and cannot be mere objects. The point at which this might actually occur remains, for Singer at least, an open question.

In a more scholarly oriented version of the argument, published in the academic journal *AI & Society*, John-Steward Gordon (2021a, 470) offers the following elucidation: "Current robots do not fully meet the morally relevant criteria (rationality, autonomy, understanding, and having social relations) necessary for them to have moral personhood and hence moral rights bestowed on them. However, we should not assume that robots will never meet these criteria; on the contrary, we should provide intelligent robots with moral and legal rights comparable to those that human beings enjoy once they have reached a certain level of functioning. At that point, it will not be up to us; rather, the apparent nature of things will compel us to grant these robots what they deserve, regardless of whether we like it." Like Singer, Gordon begins by recognizing that robots as they currently exist do not meet the necessary and sufficient condition for moral consideration or concern. Those criteria consist of a laundry list of the usual kinds of properties or capabilities: rationality, autonomy, understanding, and having social relations. But this limitation, Gordon continues, is context dependent and a matter of time. Things may be—and will likely be—different in the not-too-distant future. Consequently, robots and other seemingly intelligent and socially interactive artifacts will need to be provided with both moral and legal recognitions when they have reached a threshold condition with regards to these qualifying properties and capabilities. At this crucial tipping point, expanding the circle of moral concern so that it also includes things like robots will not be a matter of choice; the "apparent nature of things" will dictate that we cannot do otherwise.

Arguments like this—or at least proceeding according to a similar discursive strategy—are deployed by and evident in other contributions to the debate:

If we should one day succeed in creating an artificial conscious being we would have to consider it as an end-in-itself like our fellow human beings. Then we would be morally obligated to treat it as a being with intrinsic moral worth. (Göcke 2020, 239)

Adopting a broad conception of AI that includes robots, machines and other arte-facts, I argue that AIs can and should have rights—but only if they have the capacity for consciousness. (Andreotta 2021, 19)

It will be proposed that any AI that possesses the noumenal agency required to be a moral patient . . . must be granted legal personhood by any legal system that sees legal personhood as necessary for the enforcement of legal rights. (Jowitt 2021, 499)

Robots are being made with ever greater powers of cognition; at some point these powers of cognition will reach the point at which we need to start talking of robots as having minds and being persons; this will have implications for how we treat robots, for how we design robots and for how we understand ourselves and other creatures. (Reiss 2021, 68)

In all these cases—and so many others like them—the argument proceeds by mobilizing a properties approach to distinguishing between *who* has moral status from *what* does not. Properties, like consciousness, noumenal agency, cognition, or sentience, are assumed to be qualifying criteria for moral status and the ascription of rights. If robots and AI systems are capable, either now or in the future, of achieving one or more of these benchmarks, then these things not only can have rights, they should and will need to have rights. At this point, it will no longer be nonsense to advocate for the rights of robots and to say that robots or AI systems are more than mere objects. In fact, not doing so would be erroneous.

2.3   Shared Assumptions and Difficulties

Arguments like these are undeniably strong and persuasive. They issue powerful statements that have the force and effect of universal declarations or moral imperatives. And both sides in the debate have successfully mobilized this to their advantage. Alex Knapp (2011), in an article for *Forbes*, provides a rather accessible and succinct summary of the terms of the dispute:

Without the ability to make choices or think creatively beyond the bounds of its programming, an AI—no matter how intelligent-seeming—is just a big computer program. It's not a person.

Okay, but stepping into the world of speculation—let's say that we do create an artificial general intelligence that's as smart or smarter than human beings, and capable of making choices, writing poems, and all that. Would such an intelligence be worthy of respect? Almost certainly.

But I don't think it's something we'll have to worry about anytime soon, if ever.

As Knapp explains, AI systems and robots, in their current form, are nothing more than programmed artifacts or "big computers." They do not and cannot do anything beyond what they have been designed to do by their human creators. Based on this fundamental condition, AI systems and robots would not have nor need anything approaching rights because they are just objects and not moral or legal subjects. The question of rights, as well as the question of obligations, simply does not apply to these kinds of things. But if or when we succeed in creating AGI—that is, AI systems and robotic devices that are capable of actually thinking, making informed and reasoned decisions, creating original art, and so on—then they would most certainly be another kind of moral subject, one that would be worthy of respect.

Despite their differences, then, both sides agree to the same basic terms and conditions. AI systems, robots, and other artifacts are things. As long as these things remain inanimate objects lacking capabilities like conscious-ness or sentience, they are and will be just things. If and when this changes and these things possess or become capable of possessing one or more of the predesignated qualifying characteristics, then the question of moral and legal status should be on the table. Where the two sides differ is on the prob-able achievement of this final condition. The Critics argue that the time for robot rights (as Müller puts it) has not arrived and maybe never will. The Advocates assert the opposite—namely, that things will change and that we might already be seeing evidence of that change right here and right now.

Both sides therefore assert and justify their different positions by employ-ing similar argumentative strategies—placing ontological conditions or properties first (in both temporal sequence and status) and then deriving decisions concerning moral and legal standing (or the lack thereof) from these prior conditions. This is the shared understanding and common phil-osophical backdrop that both sides endorse and that make the dispute and debate between them possible in the first place. The critical problem is that these fundamental conditions—the shared set of assumptions mobilized by both sides in the debate—are themselves conditional and not at all stable, static, or settled. In fact, there are three problems with this way of proceeding that complicate the success of arguments made on either side of the dispute.

2.3.1 Determination

How does one determine which exact capability or set of capabilities are necessary and sufficient for something to have moral and/or legal status or, as Hannah Arendt (1968, 296) puts it, "the right to have rights?" In other words, which one or ones count? The history of moral philosophy and jurisprudence can, in fact, be read as something of an ongoing debate and struggle over this matter, with different capacities or psychological properties vying for attention at different times. And in this process, many criteria that at one time seemed both necessary and sufficient have turned out to be spurious, prejudicial, or both.

Take, for example, a rather brutal action recalled by naturalist Aldo Leopold (1966, 237) at the beginning of his seminal essay on environmental ethics: "When god-like Odysseus, returned from the wars in Troy, he hanged all on one rope a dozen slave-girls of his household whom he suspected of misbehavior during his absence. This hanging involved no question of propriety. The girls were property. The disposal of property was then, as now, a matter of expediency, not of right and wrong." At the time Odysseus is reported to have taken this action—something that has been recorded and preserved for us in the pages of Homer's *Odyssey*—only male heads of the household were considered legitimate moral and legal subjects. Everything else—his women, his children, his animals, his slave-girls—were property that could be disposed of without any worries or critical hesitation whatsoever. But from where we stand now, the property "male head of the household" is clearly a spurious and rather prejudicial criterion.

Similar problems are encountered with, for example, the property of rationality, which is the criterion that eventually replaces the seemingly spurious "male head of the household." When Immanuel Kant (1985, 17) defined morality as involving the rational determination of the will, non-human animals, which do not possess reason (at least since Descartes had declared that animals were nothing more than mindless extended things), are immediately and categorically excluded from consideration. The practical employment of reason does not concern animals. And when Kant does make mention of the animal or the concept of animality, he does so for instrumental reasons, only using it as a foil by which to define the limits of humanity proper. It is because the human being possesses reason that he— and *human being*, in this case, was principally defined and characterized as male, which was the "oversight" Mary Wollstonecraft sought to correct by way of her *Vindication of the Rights of Women*—is raised above the instinctual behavior of a mere brute and able to act according to the principles of pure practical reason (Kant 1985, 63).

The property of reason, however, is contested by efforts in animal rights philosophy, which begins, according to Peter Singer, with a critical response issued by English political philosopher Jeremy Bentham ([1789] 2005, 283): "The question is not, 'Can they reason?' nor, 'Can they talk?' but 'Can they suffer?'" For Singer, the morally relevant property is neither speech nor reason, which he believes set the bar for moral inclusion too high, but sentience and the capability to suffer. In his agenda-setting book *Animal Liberation* (1975) and subsequent writings, Singer argues that any sentient entity, and thus any being that can suffer, has an interest in not suffering and therefore deserves to have that interest taken into account.

Other animal rights advocates, however, dispute this determination. Tom Regan, for instance, focuses his efforts on an entirely different matter. According to Regan, the morally significant property is neither rationality nor sentience but what he calls *subject-of-a-life* (Regan 1983, 243). Following this determination, Regan argues that many animals, but not all animals (and this qualification is important, because the vast majority of animals are actually excluded from his brand of "animal rights"), are subjects-of-a-life: they have wants, preferences, beliefs, feelings, and so on, and their welfare matters to them. Although these two formulations of animal rights effectively challenge the anthropocentric tradition, there remain disagreements about which exact property or characteristic is the necessary and sufficient condition for moral consideration—that is, for some*thing* to be recognized as some*one*.

When it comes to robots and AI systems, the decisive factor—the property or capability that seems to make the difference—has been and continues to be *consciousness*. One side argues that these artifacts do not have and will likely not achieve consciousness (not even a qualitatively diminished form of consciousness, or what Ilya Sutskever of OpenAI infamously called being "slightly conscious"[6]) and therefore are and will remain mere things.

The other side argues the exact opposite. They assert that robots and AI systems, even if limited in their capabilities at this time, will at some point in the not-too-distant future attain consciousness (or something approaching what we call *consciousness*) and therefore will achieve what is necessary to be considered something more than a mere thing, becoming a moral and legal subject with both rights and responsibilities.

The seemingly irreducible problem for both sets of arguments is that the property of *consciousness* remains persistently difficult to define, characterize, and apply consistently. As psychologist Max Velmans (2000, 5) points out in his book on the subject, *consciousness* is a term that unfortunately "means many different things to many different people, and no universally agreed core meaning exists." In fact, if there is any general agreement among philosophers, psychologists, cognitive scientists, neurobiologists, AI researchers, and robotics engineers regarding the property of consciousness, it is that there is little or no agreement when it comes to defining and characterizing the concept. As roboticist Rodney Brooks (2002, 194) explains, "We have no real operational definition of consciousness," and for that reason "we are completely prescientific at this point about what consciousness is." Although consciousness, as theologian Anne Foerst remarks, is the secular and supposedly more "scientific" replacement for the occultish "soul," it turns out to be just as much an occult property (quoted in Benford and Malartre 2007, 162). Consciousness, therefore, provides a rather flimsy scaffold and not altogether solid ground for issuing and supporting decisions regarding who is to be considered a moral/legal subject and what is not.

2.3.2 Detection

Properties have problems with detection. How does one, for instance, detect the presence of consciousness in something or someone? What are the externally available signs or manifestations of its presence (or absence) in another being? How can we be reasonably certain that something that seems to possess it actually does possess it instead of merely simulating or faking it? Resolving these questions is tricky business, especially because most (if not all) of the properties and characteristics that are considered morally relevant tend to be internal mental or psychological states that are not immediately accessible or directly observable. As philosopher Paul Churchland (1999, 67) famously asked: "How does one determine whether something other than oneself—an alien creature, a sophisticated robot, a socially active computer, or even another human—is really a thinking, feeling, conscious being; rather than, for example, an unconscious automaton whose behavior arises from something other than genuine mental states?"

This is what philosophers call the *problem of other minds*. Or as Brazilian anthropologist Eduardo Viveiros de Castro (2017, 52) explains: "The theological problem of the soul of others became the philosophical puzzle of 'the problem of other minds,' which currently extends so far as to include neurotechnological inquires on human consciousness, the minds of animals, the intelligence of machines." Although the problem is not necessarily intractable, as Steve Torrance (2014) and others have argued, the fact of the matter is that we cannot, as Donna Haraway (2008, 226) effectively characterizes it, "climb into the heads of others to get the full story from the inside." Even advanced neuroimaging technology like functional magnetic resonance imaging (fMRI) does not provide an easy resolution to this epistemological obstacle. "This type of technology," as Fabio Tollon (2021, 153) explains, "allows us [to] peer into the 'moving parts' in the brain which may be correlated with sentience. However, talk of internal states and the talk of how we describe, scientifically, the information that an fMRI machine represents to us are two very different language games."

Responses to this problem typically rely on and mobilize behavioral demonstrations like that devised by Alan Turing for his imitation game, which inferred machine cognition (an internal state of mind) from a demonstration of convincing conversational behavior (an external performance).[7] Even if the behaviors are reasonably convincing, it is always a matter of inferring an internal cause from external effects. John Basl and Joseph Bowen (2020, 298) explain it this way: "We are in an epistemologically poor place when it comes to determining what the preferences of an AI are, or what makes it suffer, what it may enjoy, and so on, even if we imagine that the AI is telling us what it 'likes, enjoys, desires, etc.' and behaves accordingly. This is because whatever evidence these behaviors generate is screened off by the fact that the AI might be programmed to behave that way. Yes, the AI convincingly emotes, but it also might have been designed specifically to trick us into thinking it has mental states and emotes because of that despite having no such states." Although this phenomenon is often criticized as "deception," knowing whether it is deceit or not is part and parcel of the problem. "Deception," as Simone Natale (2021, 5) notes in his book *Deceitful Media*, "involves the use of signs or representations to convey a false or misleading impression." Therefore, to know that a particular behavior is deceptive (or not) entails that one be able to distinguish between external signs and the true situation or genuine internal state of the entity who produces or manifests those signs.

Consequently, there is, as American philosopher Daniel Dennett (1998, 172) concludes, "no proving that something that seems to have an inner life does in fact have one." Although philosophers, psychologists, cognitive scientists, neurobiologists, and the like throw an impressive amount of argumentative and experimental effort at this problem, it is not able to be resolved in any way approaching what would pass for definitive evidence, strictly speaking. In other words, no matter what property or capacity is identified—consciousness, intelligence, sentience, and so on—it is always possible to seed reasonable doubt concerning its actual presence. If an AI system or a robot, for example, appears to be sentient and therefore a subject of concern, all that is necessary to disarm this inference is to point out that it is at least possible that what appears to be intelligent behavior is in fact just an effect of clever programming or a deliberate deception.

Alan Turing attributes this rather sobering insight to a remark originally offered by Ada Augusta Byron (a.k.a. Lady Lovelace), the first computer scientist. "Our most detailed information of Babbage's Analytical Engine," Turing (1950, 450) explains, "comes from a memoir by Lady Lovelace. In it she states, 'The Analytical Engine has no pretensions to originate anything. It can only do whatever we know how to order it to perform.'" According to Lovelace, a computational device only does what we tell it to do. We can, in fact, write programs that appear to engage in interpersonal conversational behavior, like Joseph Weizenbaum's ELIZA or a digital voice assistant like Apple's Siri. This performance, however, does not mean that such a mechanism actually understands what is said to it in even a rudimentary way. Or, as John Searle demonstrated with the Chinese Room thought experiment, merely shifting linguistic tokens around in a way that looks like an understanding of language is not really an understanding of language.

Likewise, if one seeks to exclude AI systems or robots from moral consideration on the grounds that they are just things that do not possess consciousness, all that is necessary to counter this assertion is to point out how it is already complicated by available evidence. In the face of things that exhibit even rudimentary social behaviors, human beings, it seems, cannot help but engage in various forms of personification that render them more than mere things. The computers are social actors (CASA) studies undertaken by Byron Reeves and Clifford Nass (1996), for example, demonstrated that human users will accord computers and other technological artifacts social recognition similar to that of another human person and that this occurs as a product of the extrinsic social interaction, irrespective of the intrinsic properties (actually known or not) of the individual entities involved.

Social standing, in other words, is a "mindless operation." "When it comes to being social," Reeves and Nass (1996, 22) conclude, "people are built to make the conservative error: When in doubt, treat it as human. Consequently, any medium that is close enough will get human treatment, even though people know it's foolish and even though they likely will deny it afterwards." And these results have been verified in numerous "robot abuse studies," in which human-robot interaction (HRI) researchers have found that human subjects respond emotionally to robots and express empathic concern for the machines irrespective of the actual cognitive properties or inner workings of the device.

This all-too-human proclivity is often written off and dismissed as anthropomorphism. But anthropomorphism is not a bug to be eliminated; it is a feature. As Natale (2021, 132) explains: "This is as much a burden as a resource; after all, this is what makes us capable of entertaining meaningful social interactions with others. But it also makes us prone to be deceived by nonhuman interlocutors that simulate intention, intelligence, and emotions." The problem, then, is not the fact of anthropomorphism. The problem is that we fail to take it seriously as a problem. "The main problem with anthropomorphism in robotics," Kate Darling (2021, 155) writes, "is that, right now, we aren't treating it like a matter of contention. We either fall into moral panic assumptions, or we unreflectively name our robots, and if we even think twice about it, we assume it's just fun. We haven't given people's ability to relate to robots the serious consideration it will increasingly require." Consequently, even if the problem of other minds is not the intractable philosophical dilemma that is often advertised, it is sufficient for sowing reasonable doubt about how we respond to and treat things like robots and AI.

2.3.3 Decision

Finally, any decision concerning qualifying properties is necessarily a normative operation and an exercise of power. In making a determination about the criteria for inclusion, someone or some group universalizes their particular experience or situation and imposes this decision on others as the fundamental condition for moral and legal consideration. "The institution of any practice of any criterion of moral considerability," Thomas Birch (1993, 317) once wrote, "is an act of power over, and ultimately an act of violence toward, those others who turn out to fail the test of the criterion and are therefore not permitted to enjoy the membership benefits of the club of *consideranda*." In other words, every criterion of inclusion, every comprehensive list of qualifying properties, no matter how neutral, objective, or universal it appears, is an imposition of power insofar as it consists of the universalization of a particular value or set of values made by someone from a position of privilege. The problem, then, is not only with the specific property or properties that come to be selected as the universal criteria of decision but also, and perhaps more so, with the very act of decision, which already empowers someone to make and assert these defining conditions for (and all too often at the expense of) others.

Maintaining the existing boundaries is clearly about policing these decisive actions and maintaining the status quo. But the extension of rights and recognitions to those others that have been typically marginalized is no less a matter of power and privilege. As Birch (1995, 139) recognized: "The nub of the problem with granting or extending rights to others . . . is that it presupposes the existence and the maintenance of a position of power from which to do the granting." The extension of rights to previously excluded individuals or groups of individuals—although appearing to be altruistic and open to the challenges presented to us in the face of others and other forms of otherness—can only proceed on the basis of decisions instituted from a position of privilege that is more often than not the source of the exclusions that would be challenged.

This is something Mary Wollstonecraft understood and needed to negotiate in the process of crafting *A Vindication of the Rights of Women*. In order to be effective—in order to have the chance of changing anything—her argument needed to be addressed to and speak the language of those *men* who were already in a position of power, had the privilege and ability to make a change, and occupied this position precisely because of the decisive act of exclusion that was to be contested. In effect, and contrary to the famous adage issued by Audre Lorde (1984, 110), the vindication of the rights of excluded others, if it is to be legible and successful, needs to (and cannot help but) use the master's tools to tear down the master's house.

The debate concerning robot rights or the moral and legal status of artifacts confronts and has to contend with similar challenges. The Advocates are in the position of agitating for the inclusion of robots, AI systems, or other artifacts—either in general or in terms of some specific device or application—in the community of moral and legal subjects by appealing to and utilizing the very anthropocentric concepts and terminology that had been used to make and justify these exclusions in the first place. The Critics, by contrast, seem to have an easier—or at least a less burdensome—task. They only need to defend what is already standard operating procedure, using the existing privileges and power structures to support more of the same.

But because the debate is organized by and conducted in terms of rights expansion (or not), it is we human beings who are in the position of power, either to decide to grant or to deny rights claims to robots, AI systems, and other technological things. "This means," as Henrik Skaug Sætra (2021, 6) concludes, "that humans are key to determining value, as it is how entities are treated and perceived by humans that determine their moral standing." Whether the effort is situated on the side of the Critics in favor of maintaining existing ways of thinking or on the side of the Advocates, which would seek a vindication of the rights of robots, it is *we* who have granted to ourselves the right to decide who (or what) can and/or should have the right to have rights.

2.4   Outcomes and Results

Despite their many differences, both sides—the Critics and the Advocates—agree that things do not have rights or obligations. Things are objects and not subjects. And as objects, they are something to be possessed and used, not someone who has the right to have rights. The point of dispute, then, is whether robots, AI systems, and other seemingly intelligent artifacts are things or whether it is possible that they are (or could be) something more.

The Critics seek to maintain the integrity of existing decisions and ways of thinking, arguing that these devices and mechanisms—no matter how capable they are or can become—will remain things. And as things, not only do they not have and cannot have rights and obligations, but also the very question of rights and obligations simply does not apply. A phrase like *robot rights* or *AI personality* is a contradiction in terms or an oxymoron. The Advocates have a very different opinion on the matter. They concede that many of the technologies that would be called robotic or artificially intelligent—at least in the short term—are things without access to nor in need of rights or obligations. But, the argument goes, this is most likely going to change and is already in the process of changing, such that it seems prudent to consider opening up membership to the club of *consideranda*, extending social recognitions and rights to these other kinds of things that are not really things but something more, something other.

In making their case, both sides rely on and utilize the properties approach to defining the proper limits of things. The Critics argue that artifacts like robots and AI systems do not and very likely will never achieve the necessary qualifying condition of conscious, sentience, or real intelligence and that this fact justifies their continued status and treatment as objects and not subjects. The Advocates assert that these things can and will (at some point) achieve the necessary and sufficient conditions to be considered something other than mere things and therefore would be suitable candidates for being subjects of rights and obligations (or, at the very least, some limited set of rights and obligations). The difficulty for both sides is that this seemingly intuitive method has systemic difficulties with the determination and detection of the qualifying (or disqualifying) criteria.

Whether robots, AI systems, and other seemingly intelligent artifacts can be securely located in the category of thing is something that is unsettled and remains (for better or worse) an unresolved matter. On the one hand, reification of these things should be a no-brainer. They are human-made artifacts that are not conscious or intelligent in any real sense of the word and therefore should not require much effort to categorize as things. On the other hand, there is something about these artifacts, unlike so many of the other objects with which we are familiar, that seems to resist or at least significantly complicate efforts of reification. Already, right here and right now, there is something about these things that, for one reason or another, makes them other (or at least seemingly other) than mere things. But (and this is just as important) this apparent resistance to reification and containment within the category of thing does not necessarily mean that they are therefore persons. That is an entirely different, no less interesting, and equally controversial matter. And it is this question—Can or should robots be persons?—that will be taken up and developed in the next chapter.

3 Persons

The children's book *Horton Hears a Who* by Dr. Seuss (1982) famously includes the memorable and often-quoted line: "A person's a person no matter how small." The message is clear and direct. When it comes to persons, differences in what are insignificant attributes, like physical stature, should not make a difference. A person, whether they be the size of an elephant like Horton or extremely small like the almost imperceptible Whos, is equal in status and dignity. Size does not matter. It is, as St. Thomas Aquinas (2003) might say, an accident and not an essential difference.

This is also how the debate about the moral and legal status of artifacts has been organized and developed. Those who advocate for the extension of rights and obligations seek to demonstrate how apparent differences between entities—like the material of their construction—is ostensibly immaterial and how an artificially created entity can (and should) be recognized as a person. In other words, a person is a person no matter how it is made, who made it, or what it happens to be made of.

Those on the other side of the debate, by contrast, have insisted on difference—and not just apparent differences or accidents, like size or material of construction, but real differences that (so they argue) make a difference. A seemingly social, interactive, and intelligent artifact, like a robot, is definitely *not* a person, no matter how much it might look like one, act like one, or be situated in social circumstances that seem to make it one. Despite their seemingly irresolvable positions on this matter, what both sides already agree on and share is a particular investment in and understanding of what is designated by the term *person*.

3.1   What Is a Person?

Like the previous chapter's inquiry regarding things, asking the question "What is a person?" appears to be unnecessary and even superfluous. The answer is seemingly simple and straightforward: a human being. In fact, in common, everyday situations, the word *person* is often used synonymously with *human being*, and the one is easily and effortlessly substituted for the other. This intuition is not incorrect; it is just incomplete. What we need to do is track down how and why this has become the case.

The English word *person* is derived from the Latin *persona*, which originally referred to the mask worn by an actor portraying a character within the context of a stage play. In time it was extended and took on the sense of describing the guise one adopted to express certain characteristics. Only later—much later—did the word become associated with the human individual who was playing the role or taking on the guise. This evolution in terminology is, as Marcel Mauss (1985) points out, specifically Western insofar as it is shaped by the institutions of Roman law, Christian theology, and modern European philosophy. Out of this confluence of traditions we get not one but three different versions of *person*: metaphysical person, moral person, and legal person. It is as if the term *person*, like the Christian concept of the trinity, manifests itself in three distinct roles or has three separate *personae*.

3.1.1   Metaphysical and Moral Persons

In the essay "Conditions of Personhood," American philosopher Daniel Dennett begins in the usual philosophical manner, pointing out, as Heidegger had done for things, how our everyday understanding is intuitively correct but conceptually imprecise. Dennett (1998, 267) begins by recognizing the seemingly incontrovertible fact that "I am a person, and so are you" and indicates how the term is "almost coextensive" with the human being. But then he explains how the concept may extend beyond the boundaries of the human species and could be applied to other nonhuman entities. "At this time and place, human beings are the only persons we recognize, and we recognize almost all human beings as persons, but on the one hand we can easily contemplate the existence of biologically very different persons—inhabiting other planets, perhaps—and on the other hand we recognize conditions that exempt human beings from personhood" (267).

The former concerns speculation about other forms of life, which would presumably apply to nonhuman terrestrial animals. But Dennett wants to talk about space aliens. And as weird as this may sound, there is, in fact, a rather long tradition within Western philosophy concerning the moral status of extraterrestrials, including Immanuel Kant's aliens (Clark 2001) and all the talk about angels in the *Summa Theologica* of Aquinas (1945). With the latter, Dennett has in mind (using some rather dated terminology that has been the target of critical pushback in the wake of recent development in areas like disabilities studies and bioethics) "infant human beings, mentally defective human beings, and human beings declared insane by licensed psychiatrists" (Dennett 1998, 267). These human beings—these members of the species *Homo sapiens*—are often times not considered to be full moral or legal persons.

In an effort to provide a more precise and philosophically valid formulation of the term *person*, Dennett, following John Locke, untangles and distinguishes between two intertwining "notions"—one moral, the other metaphysical. A metaphysical person is "roughly, the notion of an intelligent, conscious, feeling agent." A moral person is "roughly, the notion of an agent who is accountable, who has both rights and responsibilities" (Dennett 1998, 268). The two occurrences of the adverb *roughly* are important in this context, because that word does a lot of work, indicating the extent to which many of these important and determining factors remain irregular, uneven, and undecided.

For some theorists, like medieval philosopher Boethius (1860, 1343c–d), the crucial criterion of a person is that one be an individual substance of a rational nature ("persona est rationalis naturae individua substantia"). A similar formulation can be found in the work of modern, European thinkers, like John Locke (1996, 138), for whom *person* designated "a thinking intelligent being that has reason and reflection and can consider itself as itself, the same thinking thing, in different times and places." Still others have offered more elaborate formulations that depend on a bundle of qualifying attributes and capabilities. Tom Beauchamp (1999, 310), for instance, identified five *psychological properties* that he finds deployed and operative in both classical and contemporary sources: "1) self-consciousness (of oneself as existing over time); 2) capacity to act on reasons; 3) capacity to communicate with others by command of language; 4) capacity to act freely; and 5) rationality."

Christian Smith (2010, 54), who proposes that personhood should be understood as an "emergent property," lists thirty specific capacities ranging from "conscious awareness" through "language use" and "identity formation" to "interpersonal communication and love." And Dennett (1998, 269–270), for his part, provides the following six conditions:

1. "Persons are rational beings."
2. "Persons are beings to which consciousness is attributed."
3. "Whether something counts as a person depends in some way on an attitude taken toward it, a stance adopted."
4. "The object toward which this personal stance is taken must be capable of reciprocating it in some way."
5. "Persons must be capable of verbal communication."
6. "Persons are distinguishable from other entities by being *conscious* in some special way."

The main question for Dennett is how the one category of person relates to the other, specifically whether and to what extent the metaphysical notion—the list of qualifying properties—provides necessary and sufficient conditions for being a moral subject. Despite some initial hesitation regarding this matter, Dennett (1998, 269) proceeds on the reasonable assumption that the one does in fact necessitate the other: "If we suppose there are these distinct notions, there seems every reason to believe that metaphysical personhood is a necessary condition of moral personhood." And in the remainder of the essay, he provides a detailed account of the "six familiar themes" (his words) that are typically identified as person-making properties, demonstrating how "they are necessary conditions of moral personhood" and "why it is so hard to say whether they are jointly sufficient conditions for moral personhood" (269).

If anything is certain from this arguably limited sample of different characterizations of metaphysical and moral personhood, it is that there is no univocal and agreed-upon list of criteria by which something becomes recognized as a person, and no definitive conclusion regarding whether and to what extent any list of criteria is entirely sufficient to ground moral status—that is, to decide whether something is someone or not. Dennett's inclusion of the adverb *roughly* was actually very precise and accurate. And these irregularities have no doubt provided the occasion for philosophical disagreement about these matters, but they also have an effect on the debate concerning robot rights and the moral or legal status of AI, as the two sides in the contest often mobilize different (and sometimes incompatible) criteria for deciding questions regarding who can be a person and what is not.

Despite this variability, there is one thing all of these different characterizations share and hold in common: the assumption, presupposition, or belief that the deciding factor is something that is to be found in or possessed by an individual. This is, as we have seen in the previous chapter, another iteration of the properties approach to deciding questions of moral status. You formulate a list of properties that are determined to be both necessary and sufficient for something to be considered a morally significant subject. You then go out into the world and examine whether individual things meet these criteria or not. And there are, as we previously discovered, significant challenges with the determination, definition, and detection of these qualifying properties. What has remained unacknowledged and uninvestigated, however, is the fact that this way of thinking concerns and presupposes an individual subject. In other words, it is assumed that what makes someone or something a person is some finite set of identifiable, quantifiable, and measurable *personal properties*, understood in both senses of the phrase as belonging or attributable to an individual person and as essential traits or characteristics that comprise or define what is called a person. As Charles Taylor (1985, 257) succinctly explains it, "On our normal unreflecting view, all these powers are those of an individual."[1]

This is a distinctly Western (i.e., European and Christian) way of looking at things—actually not *things*, strictly speaking, but *persons*. And this cultural specificity can be brought into focus by contrasting it to non-Western alternatives, like Ubuntu, in which *person* has been characterized otherwise.[2] In these traditions (and it should be noted that this is not one univocal tradition but a constellation of different but related traditions), *person* is not understood as the natural condition of an individual entity, who it is assumed is a person due to some predetermined set of definitive properties. It is an achieved social condition or recognition. Theorized in this way, the title of *person* is not something naturally bestowed on or belonging to an individual; it is "something which has to be achieved" through a social process and due to recognition by others (Menkiti 1984, 172). This alternative formulation is not entirely alien to and unknown within Western traditions. It is just situated and developed under a different name—that is, *legal person.*

3.1.2 Legal Person

A *legal person* is distinguished from the moral and metaphysical concept of a person to the extent that it focuses attention not on what something is in its essence—that is, a laundry list of metaphysical, person-making qualities or capabilities—but on how it is situated in real-world, external relationships with others. The term *legal person,* as Ngaire Naffine (2009) explains, "is strictly a formal and neutral legal device for enabling a being or entity to act in law, to acquire what is known as a 'legal personality': the ability to bear rights and duties."

This formal and neutral aspect has, as John-Stewart Gordon (2021a) explains, two important consequences, both of which should sound familiar insofar as they were already mobilized by Dennett in "Conditions of Personhood." First, not all human beings are persons. Although adult human beings are considered to be both moral and legal persons, others are not. "Children, new-borns, people with severe mental impairments, and people in a non-responsive state . . . may carry some degree of moral status, but do not have the same moral and legal personhood as that granted to a typically functioning adult human being" (Gordon 2021a, 457). Still others, like convicted criminals, might retain full moral personhood but be denied some of the rights and protections accorded legal persons, such as the right to vote or to move freely without restriction.

Second, not all persons are human beings. There are numerous other entities—artificial, like corporations, and naturally occurring, like the things of nature—that are recognized as legal persons but that may not be moral persons. "Whether corporations and trust funds, which are commonly considered *legal persons,* also have *moral personhood* is debated in the realms of legal philosophy and law." Likewise, "one might contend that some environmental objects, such as holy rivers and unique national parks, have a high moral status and, therefore, should also be treated as legal persons. But to argue that they could also qualify as moral persons simply because they are considered legal persons in some jurisdictions seems farfetched" (Gordon 2021a, 457).

Characterizations of the term *legal person* have evolved over time and within context of use. Initially, as explained by legal historian J.-R. Trahan

(2008, 10), "the Roman jurisconsults seems to have taken the concept to include, first and most fundamentally, a 'human being' or, better yet, every human being properly so called, including slaves." This characterization sounds good in theory, but it was exclusive in practice. For the Romans, there was a fundamental difference between "persons of their own right" (*personae sui iuris*) and slaves, which were characterized as "persons subject to another's right" (*personae alieno iuris subiectae*; Spaemann 2006, 23). Consequently, what we typically regard and recognize as a *person* only applied to and designated the *paterfamilias*; it did not concern his women, children, or slaves. For this reason, person was more of a social designation (the role that one plays or occupies) than an ontological category (what one is). "Because the *persona* was never identical to the face," Esposito (2015, 30) writes, "the *persona* was not the individual as such, but only its legal status, which varied on the basis of its power relationships with others."

Others in this context refers not only to other persons but also and principally to things. In fact, things play a crucial role here, because *person* not only designates the opposite of *thing* but also was characterized in terms of the possession of and dominance over things. "Since a thing," Esposito (2015, 17) explains, "is what belongs to a person, then whoever possesses things enjoys the status of personhood and can exert his or her mastery over them." This means that the fundamental conceptual opposition instituted between person and thing is not and never was neutral or what Derrida (1981, 41; emphasis in original) calls the "peaceful coexistence of a *vis-à-vis*." The two terms are not on equal footing; one already has dominance and privilege over the other. "Persons own things, and things are owned by persons" (Iwai 1999, 587).

Modern characterizations extend and modify this fundamental understanding. "Regarding 'persons,'" Trahan (2008, 12) reports, "Grotius added little to the stock of existing ideas, but what little he did add proved to be important: 'persons,' he wrote, are those who 'have rights to things.' Though Grotius himself did not say as much, this attribute of persons clearly implies—indeed, presupposes—another, namely, that persons 'can' have such rights, in other words, have the 'capacity' to receive or acquire them." This "new take" (Trahan's phrase) on the concepts of person and thing—one that is organized in terms of and proceeds according to rights—is further developed and refined in Anton Thibaut's *An Introduction to the Study of Jurisprudence* (1855, §101, at 88): "We have next to consider the subjects of rights and duties, that is to say, the persons to whom something is possible or necessary. In the first place we must examine who or what, either from its very nature or by the precepts of positive law, can be considered as capable of rights and duties. By Person is meant whatever in any respect is regarded as the subject of a right: by Thing, on the other hand, is denoted whatever is opposed to person."

Characterized in this fashion, *person* is not, as Jenny Tiechman (1985, 184) concludes, "the name of a natural species; nor is it the name of a broad natural kind." It identifies the subject of rights and duties. "So far as legal theory is concerned," John Salmond (1907, 275) concludes, "a person is any being whom the law regards as capable of rights and duties. Any being that is so capable is a person, whether a human being or not, and no being that is not so capable is a person, even though he be a man. Persons are the substances of which rights and duties are the attributes."

3.2   What Are Rights?

If persons—and legal persons in particular—are understood as the subject of rights and duties, then what do *rights* and its correlate *duties* or *obligations* designate? Like *time* in *The Confessions of St. Augustine*, *rights* is one of those words that we are all fairly certain we know the meaning of, up to the point that someone asks us to define it. Then we run into difficulties and confusions. This is neither unexpected nor uncommon.[3] One hundred years ago, an American jurist, Wesley Hohfeld, observed that even experienced legal professionals tend to misunderstand the term, often using contradictory or insufficient formulations in the course of a decision or even a single sentence.

So let's start with a standard dictionary definition. "Rights," as Leif Wenar (2020) explains, "are entitlements (not) to perform certain actions, or (not) to be in certain states; or entitlements that others (not) perform certain actions or (not) be in certain states." This characterization is technically accurate but not very portable or immediately accessible. To get a better handle on the concept and how rights actually work, we can break it down following Hohfeld's pioneering work in this domain.

In response to what he perceived to be confusions regarding the (mis)use of the concept, Hohfeld (1920) developed a typology that analyzes rights via four molecular components, or what he called *incidents*: claims, powers,

Figure 3.1
Hohfeldian analysis of the right to property ownership for a toaster. Original image by the author.

privileges, and immunities. His point was simple and direct: a right, like the right one has over a piece of property, can be defined and operationalized by one or more of these incidents. It can, for instance, be formulated as a claim that an owner has over and against another individual. Or it could be formulated as an exclusive privilege for use and possession that is granted to the owner of the object. Or it could be a combination of these (figure 3.1).

Hohfeld also recognized that rights are fundamentally social and relational. The four types of rights or incidents only make sense to the extent that each one necessitates a correlative duty that is imposed on at least one other individual. "The 'currency' of rights," as Johannes Marx and Christine Tiefensee (2015, 71) explain, "would not be of much value if rights did not impose any constraints on the actions of others. Rather, for rights to be effective they must be linked with correlated duties." Hohfeld, therefore, presents and describes the four incidents in terms of rights/duties pairs:

If A has a Privilege, then someone (B) has a No-claim.

If A has a Claim, then someone (B) has a Duty.

If A has a Power, then someone (B) has a Liability.

If A has an Immunity, then someone (B) has a Disability.

This means that a right—like a claim to property ownership—means little or nothing if there is not, at the same time, some other individual who is obligated to respect this claim. Or as Jacob Turner (2019, 135) explains, mobilizing an example that finds expression in both European literature and philosophical thought experiments: "It would not make sense for a person marooned alone on a desert island to claim that she has a right to life, because there is no one else against which she can claim that right." On Hohfeld's account, then, rights are a social phenomenon. A solitary human being living alone without any contact with another person (something that is arguably a fiction) would have no need for rights.

Furthermore, and as a direct consequence of this, rights can be perceived and formulated either from the side of the possessor of the right (the power, privilege, claim, or immunity that one has or is endowed with), which is a "patient-oriented" way of looking at a moral, legal, or social/political situation, or from the side of the agent (what obligations are imposed on another who stands in relationship to this individual), which considers the responsibilities or obligations of the producer of a moral, legal, or social/political action. The debate about robot rights, then, is not just about robots; it is also and inextricably about us—we who would be obligated by and responsible for responding to the claims, powers, privileges, and/or immunities belonging or assigned to the robot, AI, or other artifact.

3.3 Having Rights

But how does a person come to have or be assigned rights? What justifies an entity possessing rights or not? How does one become subject to or the subject of rights? As explained by Thibaut (1855, §101, at 88), a person has or is bestowed with rights "either from its very nature or by the precepts of positive law." The former, what is called *natural law* (*ius natural* or *lex naturalis*), is based on and derived from properties or qualities that are determined to be intrinsic to something's essential nature. *Positive law* (*ius positivum* or *lex posita*), by contrast, is socially constructed or posited (the words *positive* and *posit* share a common etymological root, meaning "to place") by explicit acts of legislation. In other words, the specifications of natural law are considered to be true by nature, whereas positive laws are true by declaration and the exercise of human decision-making and exertions of power. Distinguishing between these two concepts is crucial for a number of reasons, not the least of which is that participants in debates about these matters often "shift between moral and legal frames without fully appreciating how they differ in terms of the criteria applied and the conclusions they reach as a result" (Gellers 2020, 28).

3.3.1 Natural Rights

"All natural rights theories," as Wenar (2020) explains, "fix upon features that humans have by their nature, which make respect for certain rights appropriate. The theories differ over precisely which attributes of humans give rise to rights." In religious traditions, this is something that is typically explained and justified by appeal to divine or transcendental authority. In Christianity, for instance, the "rights of man" (and the gender-exclusive formulation is deliberate in this context) are justified by the doctrine of the *Imago Dei*, the belief or *doxa* that human beings—beginning with the first man, Adam—have been created in the image of god[4] and bestowed by their creator with inalienable rights. In nonreligious or secular traditions, the determining factors are "the same sorts of attributes described in more or less metaphysical or moralized terms: free will, rationality, autonomy, or the ability to regulate one's life in accordance with one's chosen conception of the good life" (Wenar 2020).

Natural or what are also called *moral rights*—mainly because these are the rights belonging by nature to a moral person—provide a strong case for inalienable, universal entitlements and protections. The Declaration of Independence, issued in Philadelphia on July 4, 1776, for example, begins with a forceful natural rights assertion: "We hold these truths to be self-evident, that all men are created equal, that they are endowed by their Creator with certain unalienable Rights, that among these are Life, Liberty and the pursuit of Happiness." A similar assertion is made in the first lines of the *Déclaration des droits de l'homme et du citoyen de 1789* [Declaration of the Rights of the Man and of the Citizen of 1789] sans the appeal to divine authority: "Men are born and remain free and equal in rights. Social distinctions may be founded only upon the general good. The aim of all political association is the preservation of the natural and imprescriptible rights of man. These rights are liberty, property, security, and resistance to oppression." Likewise for the UN Declaration of Human Rights, which begins with a similar kind of universal proclamation grounded in and guaranteed by the essential nature of the human being: "All human beings are born free and equal in dignity and rights. They are endowed with reason and con-science and should act towards one another in a spirit of brotherhood."

Natural or moral rights statements like these are strong assertions of fundamental claims, powers, privileges, or immunities that are declared to belong to every man or human being (and the gender-exclusive formula-tion that is made in many of these historically important statements is not without its own problematic history) due to their very nature as human beings. The ultimate justification and guarantee of the truth of these uni-versal declarations is something that is predicated on and referred to a tran-scendental or universal authority, either a divine creator or the natural state of the human being, what is often called (within European philosophical traditions) *human nature*. This is simultaneously the source of immense dis-cursive power and systemic weakness. What makes these naturalistic for-mulations persuasive, forceful, and seemingly indisputable is that they are grounded in and guaranteed by a transcendental, universal authority that stands outside and beyond specific, finite human constructs, institutions, and expressions of power.

But this is also what makes these statements fragile because the exis-tence of this universalized transcendental underwriter is more often than not a matter of faith or philosophical speculation and not a scientifically proven fact. This is one of the reasons that in the history of philosophy (at least in its European configuration), a proof for the existence of god has typically been pivotal and necessary. The system only works and is able to stand on its own if the existence of the transcendental authority anchor-ing and propping up the entire edifice can itself be anchored and guar-anteed. This was one of the important and enduring insights of Friedrich Nietzsche (1974): If it is the case that "god is dead," then everything that had been supported and guaranteed by his authority is rendered vulnerable and potentially meaningless.

3.3.2 Legal Rights

The other way persons have rights is through the precepts of positive law. According to this formulation, rights are not justified by and derived from the essential nature of the rights holder and propped up by an appeal to transcendental authority: instead they are conventional rules or socially constructed stipulations. As Turner (2019, 135–136) explains, "Rights are collective fictions, or as Harari (2015) calls them 'myths.' Their form can be shaped to any given context. Certainly, some rights are treated as more valuable than others, and belief in them may be more widely shared, but there is no set quota of rights which prevents new ones from being created and old ones from falling into abeyance." This is both good news and bad news.

First the good news. Unlike natural rights, legal rights do not depend on fanciful metaphysical speculations about the essential nature of things nor appeal to supernatural authorities, the existence of which can always be doubted or questioned. But—and here's the bad news—this means that legal rights are a matter of human (all too human) decision-making and that the assignment, distribution, and protections of rights are ultimately a matter of finite exercises of terrestrial power. Where natural rights are anchored in eternal metaphysical truths that can be discussed and debated by theologians and philosophers, legal rights are legitimated by earthy exercises of specific sociopolitical power.

For this reason, legal justifications for rights are considered to be "weaker." Because they are ultimately anchored in and legitimated by conventional agreement, they are not only alterable (i.e., able to be modified, repealed, and restrained) but relative, meaning that they exhibit significant variability across different human communities distributed in time and space. Saying this is not, as Turner (2019, 136) is quick to point out, a critique; it is descriptive and value neutral. "Describing rights as fictions or constructs is by no means pejorative; when used in this context, it does not entail duplicity or error. It simply means that they are malleable and can be shaped according to new circumstances." And technological innovations, like robots, AI systems, and other seemingly intelligent artifacts, certainly provide ample opportunities and challenges for "new circumstances."

3.4 Natural versus Artificial Persons

Like rights, which can be differentiated into natural and legal categories, there is a similar binary division affecting the concept person, where we typically distinguish between natural persons and artificial persons. This difference, as Visa A. J. Kurki (2019, 6–7) explains, is originally rooted in modern European legal innovations and subsequently exported and extended beyond that initial context:

> Theories of law, of rights, and of legal personhood that were reached in nineteenth-century Germany would have a very profound impact on certain core tenets and

taxonomies of legal personhood, which are still endorsed by jurists across the Western world. These include . . . the fundamental divide into 'natural persons' (*natürliche Personen; personnes physiques*)—denoting human individuals who are legal persons—and 'artificial' or 'juristic persons' (*juristische Personen; personnes morales*), meaning any other types of legal persons, such as associations, limited liability companies, and foundations, all of which can own property and enter into contracts in their own names.

Person divides into two types. There are natural persons, which typically are human individuals or other entities who can be considered persons by nature. And there are artificial or juristic persons, who are persons not by their nature but by an act of law or official decree. Remixing Dr. Seuss, we can say that a person is a person, no matter whether naturally occurring or artificially designated.

But this is also where terminological distinctions become messy and confused. In the legal literature, there is a fundamental division between natural and artificial persons. "The most common and intuitive definition of a legal person," as Alexis Dyschkant (2015, 2078) explains, "is a natural person or human being. Humans are called 'natural' persons because they are persons in virtue of being born, and not by legal decree. Corporations and even inanimate objects, such as ships, have been deemed legal persons by decree, and thus are non-natural."[5] But as we have already seen, the philosophical literature tends to distinguish between two (sometimes three) different kinds of persons: moral persons and legal persons. The former is determined and defined by a set of metaphysical, person-making properties; the latter is an artifact of human social institutions. As a result, the use of terminology in the debate about the moral and legal status of robots, AI systems, and other artifacts is not exactly precise, and there are often unacknowledged substitutions, and even confusion not just between different publications but also (in some cases) even within the space of the same publication.

For example, the term *legal person*, as Kurki (2019, 7) points out, is often substituted and used as a synonym for nonnatural or artificial persons, mainly because in these instances and situations an entity is defined as a person not due to its innate natural qualities but by legislative decree or judicial decision. But properly speaking, *legal person* designates who the law considers to be the subject of rights and duties, and those entities may be either natural persons, like a human being, or artificial persons, like

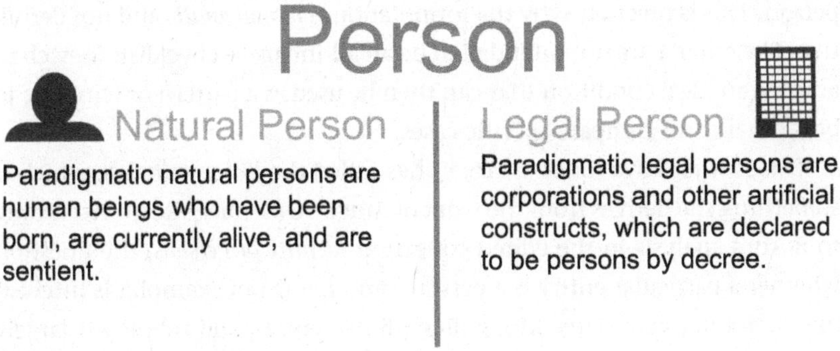

Person

 Natural Person | Legal Person

Paradigmatic natural persons are human beings who have been born, are currently alive, and are sentient. | Paradigmatic legal persons are corporations and other artificial constructs, which are declared to be persons by decree.

Figure 3.2
Natural versus legal person. Original image by the author.

corporations, organizations, ships, and robots. Likewise, the term *moral person* is sometimes utilized as another name for what is designated by *natural person*. But again, something artificial, like a robot, could be a moral person, if it was able to model the right set of person-making qualities that are said to belong to a paradigmatic natural person like a human being. We are not (even if we wanted to) going to be able to sort this out once and for all and to the satisfaction of everyone involved in the debate. What we can do and need to do is to learn to be sensitive to these different sets of terminological distinctions so that, in the course of pursuing the analysis, we do not mistake semantic differences in vocabulary as fundamental disagreements about the subject matter being discussed and debated (figure 3.2).

3.4.1 Natural Persons
Natural persons have typically been defined by way of identifying a paradigmatic instance or example, which then serves as an ideal form or template for anything that would be considered a person by nature. "In contemporary Western legal systems," Kurki (2019, 8) explains, "the paradigmatic natural persons are (1) human beings, (2) who have been born, (3) who are currently alive, and (4) who are sentient." If one takes this list as definitive—meaning that *person* would only apply in situations where the entity in question was able to meet all four conditions—it would be difficult to see how anything other than a "normal human being" (a formulation that is not without controversy and rightfully critiqued in both biomedical ethics and disability studies) would be able to be considered a

person. This is precisely why the formulation is *paradigmatic* and not definitive. The criteria are not intended to be an all-inclusive checklist; they characterize an ideal condition that can then be used as a pattern or template to benchmark and compare specific cases.

This is what Raya Jones (2016, 9) has called the *like-us criterion*, which is something she derives from the work of Amélie Oksenberg Rorty. According to Rorty's analysis in the edited collection *Identity of Persons*, the question whether a particular entity is a person—and her list of examples is interesting, including Venusians, Mongolian idiots, fetuses, and robots—is largely a conceptual one that depends on our ability to determine to what extent "things look like us": "If Venusians and robots," Rorty (1976, 322) writes, "come to be thought of as persons, at least part of the argument that will establish them will be that . . . while they are not the same organism that we are, they are in the appropriate sense the same type of organism or entity."

This is precisely the mode of argument that has been deployed and put into practice by animal rights philosophers, like Peter Singer and Tom Regan. According to Regan (1983, 76), the case for animal rights (the title of his seminal book on the subject) does not include all animals but is limited to those species with sufficient complexity to have at least a minimal level of abilities similar to a human being: "The greater the anatomical and physiological similarity between given animals and paradigmatic conscious beings (i.e., normal, developed human beings), the stronger our reasons are for viewing these animals as being like us in having the material basis for consciousness; the less like us a given animal is in this respect, the less reason we have for viewing them as having a mental life."

For Regan and other personists, animal rights are intentionally restricted. Not all animals are equal; some animals are more equal than others. And what makes the difference is the extent to which a particular animal or kind of animal "looks like us" by approximating the abilities of a "a normal, developed human being" (once again, the word *normal* as it is used in this context is not without critical difficulties proceeding from ableist assumptions). Consequently, the task we face in the face or faceplate of the robot, AI system, or other artifact is not to prove that these technological devices are exactly equivalent to a naturally occurring human being but that they are—like some animals—capable of achieving or giving evidence of the benchmarks of a paradigmatic natural person. This means that the decision will need to be concerned with both ontological matters—that is, determining the criteria or qualifying properties that make something a person—and epistemological issues—that is, detecting the presence (or the absence) of these properties in another entity.

In addition to managing these two variables, there is also an underlying political problem. With decisions regarding natural persons, human beings are (or, perhaps stated more accurately, have already granted to themselves the power and privilege to occupy the role of) both measure and measurer. Human beings not only furnish the standard template for what is a paradigmatic natural person but are the final word on who or what meets those criteria or not. We are, then, both judge and jury. There are two problems with this.

First, there is the potential for a conflict of interest—the fact that *we* define the criteria and then decide who or what meets these standards or not. This has allowed us (or stated more accurately, provided some of us the privilege to allow ourselves) to modify, change, or even suspend the rules of the game when it serves our own interests. And a good illustration of this can be seen in what has been called the *AI effect* (Kaplan 2016, 37). Consider the game of chess, which is not just one example among others but has been one of the principal defining conditions in the field. For decades, the task of playing championship-level chess was seen as a challenge that would require the achievement of human-level intelligence to resolve. As Claude Shannon (1950, 257) predicted in his field-defining paper on the subject: "Chess is generally considered to require 'thinking' for skillful play; a solution of this problem will force us either to admit the possibility of a mechanized thinking or to further restrict our concept of 'thinking.'" And once the objective had been achieved—in 1997, when IBM's Deep Blue defeated the reigning human champion, Garry Kasparov—playing championship-level chess became just another computer application and was no longer considered a qualifying benchmark for intelligence. As soon as the computer reached the goal line, we simply moved the goal posts and redefined what was necessary for something to be considered intelligent.

Second, there is a prior and often unacknowledged decision regarding who is included in the first-person-plural pronoun *we* and who/what remains excluded and marginalized in the process. Decisions regarding natural persons proceed by allowing (or empowering) a specific group of human beings to decide what is natural and to grant to themselves the unique privilege to speak on behalf of the universality of nature. This specific exercise of power has historically resulted in troubling outcomes, not only permitting some groups of human beings to normalize their own experiences at the expense of others but also instituting asymmetrical relationships of power, in which *we* has been an exclusive moniker, marginalizing the insights and potential contributions of others.

Consequently, *natural person* is anything but natural. It is a conceptual fabrication belonging to a specific culture and therefore an artifact of its intellectual traditions and social institutions. It is an instance of what Donna Haraway (1991) has called *situated knowledge*. When the word *we* is deployed, one must always ask: Who are we? Who speaks for us? And ultimately, what particular interests and investments are being served by this seemingly universal designation? Unfortunately, we often do not know to what extent *we* is already a critical problem, instituting exclusive decisions that marginalize others and stack the deck in our favor.

3.4.2 Artificial Persons

Artificial persons, by contrast, are determined to be persons not because of their intrinsic nature but by way of an external decree. Something is a person not because of its (so-called) natural characteristics, but because someone says that it is and has the power to authorize and enforce that declaration. To be a person, then, means that one is recognized as a subject under the law, possessing both responsibilities and rights within a particular legal construct or jurisdiction. If the human being has been the paradigmatic natural person, then the paradigmatic artificial person (or what is also called a *legal person*) is the corporation (Dyschkant 2015, 2084).

Corporations are persons with rights and responsibilities similar to those of a natural person. But unlike a natural person, they have this status not due to a set of intrinsic natural properties—consciousness, sentience, being alive, and so on—but by being recognized and situated within the law as subjects of the law. As Dyschkant (2015, 2085) explains: "While human beings are born as persons, such that the law cannot help but recognize them as persons, corporations are merely a legal construct." Or as explained in the 1819 US Supreme Court decision regarding the landmark case of *Dartmouth College v. Woodward* (17 U.S. 518, 636): "A corporation is an artificial being, invisible, intangible, and existing only in contemplation of law. Being the mere creature of law, it possesses only those properties which the charter of its creation confers upon it, either expressly, or as incidental to its very existence." Although an artifact like this may be regarded as little more than a legal fiction (*persona ficta*), it is a fiction with very real consequences.

Formulated in this way, the concept of person is a socially constructed status that is conferred on others by those in a recognized or mutually agreed-upon position of power. *Person*, in this context, designates not what one is by way of their nature or essence; it identifies the position one occupies or the role that one plays in social relations with others. What matters then are not ephemeral metaphysical qualities and/or statements regarding something's essential nature, but real-world outcomes and social utility. "The corporation," as Katsuhito Iwai (1999, 590) explains in his essay on the subject, "is understood here primarily as a legal device which simplifies and stabilizes the complicated web of contractual relationships that an association of shareholders has to have with a multitude of outside parties. Its legal personality endows the corporation with the legal capacity to interpose itself between shareholders and outside parties and to enter into contracts with the latter on behalf of the former." This has two important consequences that directly impact the debate about robot rights and the moral and legal status of AI systems.

First, the question whether an entity, like a robot, AI system, or other artifact, could be an artificial person is less about the entity and its specific capabilities and more about us and our social institutions. And it will help in this situation to proceed as my previous book, *Robot Rights*, did, by dividing the "can" aspect of the question from the "should." If we ask the question, "Can robots be persons under the law?" the answer is unequivocally "yes," but not because of what the robot is (or is not). It is because of the way law works. All that is necessary for something to be recognized as a person is for some legal authority—the head of state, a legislature, or a court of law—to decide that, for whatever reason, some robot or AI system (or even a class of robots or AI systems) has status as a person. This is how corporations became recognized as artificial persons. This is why animal rights activists, like Nonhuman Rights Project, and environmental groups are petitioning courts to officially recognize the legal status and rights of animals, mountains, and waterways. And this is why, in a recent proposal issued to the European Parliament by the Committee on Legal Affairs (Delvaux 2016, 12), it was suggested that AI systems and robots might need to be considered "electronic persons" for the purposes of legal integration and tax policy.

But just because you can, as David Hume knew and correctly pointed out, does not mean that you should. Just because a legislature or court can bestow what is often called *legal personality* on an artifact does not mean that it should be done or that doing so is a good idea. The ability to do something like this does not automatically mean that it is correct or socially responsible to do so. This is where things diverge, with each side in the debate providing diametrically opposing responses to the question, "Should robots be artificial (or legal) persons?" The Critics recognize that it is both possible and entirely logical for robots, AI systems, and other artifacts to be granted legal personality and be positioned in law as artificial persons similar to corporations, but they argue that doing so would be a very bad idea, if not catastrophic. The Advocates, by contrast, argue that the social circumstances of the twenty-first century make it necessary that we devise ways to integrate all kinds of nonhuman entities into our legal systems and that extending the status of person to robots, AI systems, and other artifacts will be necessary to maintaining the integrity of our social institutions.

Second, because legal personality is a matter of decree, these efforts do not need to get into the weeds regarding the big metaphysical questions, like the nature of intelligence, the qualities of consciousness, the meaning of the doctrine of the *Imago Dei*, or what set of unique person-making prop- erties are necessary and sufficient for something to be considered someone. Instead, they are concerned with more pragmatic and down-to-earth issues. A robot or an AI system, like a corporation, will be a person if and when it is decided and declared by law to be a person. And what ultimately matters in making this decision is not a set of ontological properties belonging to the technological system or device, but the very real effect and impact the decision will have on existing social structures and institutions. This appar- ent advantage, however, is not without its own complications and costs.

Legal declarations and decisions are bounded by jurisdiction. Conse- quently, what is declared to be a person in one domain may not be recog- nized (or not recognized in the same way) in another. We can already see how this works and what it means by considering recent efforts to extend the recognitions of personhood to natural objects of the environment. In 2017, New Zealand declared the Whanganui River to be a legal person.[6] This new legal entity is identified with the proper name Te Awa Tupua and is recognized as "an indivisible and living whole from the mountains to the sea, incorporating the Whanganui River and all of its physical and metaphysical elements" (New Zealand Parliament 2017). And as is typically done in corporate law, the river has been appointed two guardians to represent it and speak on its behalf. But extending the legal status of person to this one river in this one location does not mean that other rivers in other regions of New Zealand or other parts of the world are also persons. Each case is specific, limited, and context dependent.

3.5 · Outcomes and Results

The main advantage of artificial persons also turns out to be its principal liability. On the one hand, decisions regarding artificial personhood are far more pragmatic and practical than those involving natural persons. Whereas the latter needs to get into the thick of things with ontological properties and the epistemological complications of measuring their presence or absence in the mind or body of another entity, artificial personhood is simply decided and assigned. It's not about obscure metaphysical causes or psychological qualities; it is about very real social outcomes and effects.

But this means, on the other hand, that designations of personhood can be criticized for being capricious, will always be context specific, and may even result in disquieting inconsistencies. As an example, one only needs to consider the storied and contentious history of corporate personhood as it has played out and been debated in the past several decades. Unlike natural personhood, which anchors decisions and declarations in seemingly universal criteria that are true by nature, the artificial person status is a socially constructed artifice, lacking that kind of firm and certain metaphysical foundation. Ultimately, this means that the arguments both for and against robot rights and AI personhood will be organized and proceed very differently based on differences in the kind of person that comes to be deployed and operationalized.

The next two chapters will address this. The first one (chapter 4) will investigate whether and to what extent robots, AI systems, and other seemingly intelligent and socially interactive artifacts might need to be considered natural persons. The second one (chapter 5) will do the same for the concept of legal or artificial personhood. Splitting the examination into two separate chapters may appear (at least initially) to be "too much." But the wide range and depth of available research material and prior publications regarding this subject make it a practical necessity.

4 Natural Persons

The opportunities and challenges of using the concept of natural person in debates about the moral and legal status of robots, AI systems, and other seemingly intelligent artifacts is something that has been insightfully portrayed and prototyped in Justin Leiber's *Can Animals and Machines Be Persons?* This fictional "dialogue about the notion of a person," as Leiber (1985, ix) characterizes it, consists of the imagined transcript of a hearing before the United Nations Space Administration Commission (a fictional UN commission) and concerns the fate of two nonhuman inhabitants of a space station: a young female chimpanzee and the station's AI computer. The dialogue begins in medias res. The space station is beginning to fail and needs to be shut down and decommissioned. Unfortunately, doing so means terminating the life of both its animal and machine occupants, which, according to stipulations provided in the narrative, cannot be suitably evacuated.[1] In response to this news, nonhuman rights activists have opposed the decision, arguing that the shutdown would violate the rights of these two nonhuman persons.

Leiber's dialogue, therefore, takes the form of a debate between two parties: (1) a complainant who argues that the chimpanzee and computer, though not human, have the capacity to think and feel, and therefore should be considered persons with the same rights and duties accorded a human person; and (2) a respondent who asserts the opposite—namely, that neither of these two things are persons because only "a human being is a person and a person is a human being" (1985, 6). What is important in this "naturalistic discussion" (and these are the words Leiber uses to describe it) is not just the terms and conditions of the debate, which are reproduced with remarkable fidelity in the subsequent nonfiction literature, but the fact that the benchmark for person used by both sides in the debate is a human being, and the judge and jury in the case are also human beings. In any other circumstance, this would be cited for, or at least identified as, a potential conflict of interest. And as we shall discover, this complication generates its own set of consequences.

4.1 The Critics and Their Arguments

Asserting that robots, AI systems, or other artifacts cannot be natural persons appears to be intuitively correct and noncontroversial. Natural persons are in both name and definition *natural*; they are persons by their very nature. They have been born, are currently alive, and are (self-)conscious or in possession of sentience. Robots, AI systems, and other kinds of things are—again, by name and definition—*artificial*. They are designed, manufactured, and deployed by human persons. As Joanna Bryson (2010, 65) correctly asserts, "There would be no robots on this planet if it weren't for deliberate human decisions to create them." Robots are, in other words (specifically, the words of Greek metaphysics), the product of τέχνη (*tékhnē*) and not φύσις (*phúsis*).[2] They are artificially made and not naturally occurring. They may be designed and constructed to model, emulate, and simulate the look, feel, and behavior of a naturally occurring entity, like a human being or a nonhuman animal, but they remain artificially constructed objects that are not really what they seem to be.

Although this appears to be rather obvious and noncontroversial, the devil is in the details. In an early publication on the subject, cognitive scientist Steve Torrance identified and analyzed a mode of argument he called the *organic view*. According to Torrance, who does not himself espouse the position but merely seeks to formulate its logical structure, the organic view includes the following five related propositions:

1. There is a crucial dichotomy between beings that possess organic or biological characteristics, on the one hand, and "mere" machines on the other.

2. It is appropriate to consider only a genuine organism . . . as being a candidate for intrinsic moral status—so that nothing that is clearly on the machine side of the machine-organism divide can coherently be considered as having any intrinsic moral status.

3. Moral thinking, feeling and action arises *organically* out of the biological history of the human species and perhaps many more primitive species which

may have certain forms of moral status, at least in prototypical or embryonic form.

4. Only beings, which are capable of sentient feeling or phenomenal awareness could be genuine subjects of *either* moral concern *or* moral appraisal.

5. Only biological organisms have the ability to be genuinely sentient or conscious (Torrance 2008, 502–503).

Formulated in this way, the organic view draws a clear line of demarcation, instituting a binary distinction that neatly divides one category of entity from another. On the one side, there are living things—organic or biological organisms that are sentient and therefore legitimate subjects of moral consideration and rights attribution. These are, though Torrance does not use the term, natural persons. On the other side, there are mere mechanisms—nonliving things that have no proper claim to moral concern or appraisal whatsoever. These things are just things. And these ways of organizing things, as Torrance (2008, 503) rightfully concludes, "definitely exclude robots from having full moral status."

4.1.1 Organic View Arguments

Many of the subsequent efforts conform to the basic elements and parameters of this formulation, even if they do not self-identify as instances of what Torrance calls the organic view. Consider, for example, the argument provided by Kathleen Richardson (2019), who not only emphasizes the essential ontological differences between living animals and machines but also criticizes the proponents of robot rights for incoherently and incorrectly confusing the one with the other:

> The proponents of "robot rights" or "machines rights" map onto machines, human relations with animals, which are wholly incoherent. Machines are always property; they are made up of parts that are commodities. Animals may be exchanged as commodities but their bodies and lived experiences exist outside of human-made artefacts. Animals are pulled into property relations of human cultures, but they are not it. Robots and AI can never exist outside of property relations because they are commodities. Robots and AI are inside property relations, not living relations. Robots and AI at best might mimic behaviours and relations, but they are not participating as living and sexed relational beings. (Richardson 2019, 80)

Richardson's argument turns on an essential, mutually exclusive distinction that divides living beings from what are mere things and commodities (which is a restatement of the first proposition listed in the organic view). The former, she asserts, rightfully participate in social relationships that involve matters of rights and duties, while the latter simply and by nature do not. Thus, had Richardson been party to Leiber's fictional debate, she would argue that we need to split the difference: that is, recognize the chimpanzee as a genuine organism, participating in life as a living and sexed relational being, and therefore a legitimate candidate for rights and protections, while summarily dismissing the claim of the AI to any such status because, unlike the chimpanzee, it is just an artificial thing and not a natural living being.

Another iteration of the argument is supplied in Adam J. Andreotta's "The Hard Problem of AI Rights." For Andreotta, a rights claim for AI—which in this case designates a wide range of technological things, like "artificially intelligent robots, machines, and other artefacts" (Andreotta 2021, 19)—is possible if and only if we can first resolve the *hard problem of consciousness*, a term and concept that he takes from the work of David Chalmers (2010, 8). But as we cannot solve for this, AI will not be and cannot be the subject of rights:

> I argue that AIs can and should have rights—but only if they have the capacity for consciousness. This mirrors the same reasoning that is commonly employed in discussions about animal rights. Problematically, for AI rights, however, the analogy with animal rights is not perfect. Since we share an evolutionary history with mammals, birds, reptiles, and other nonhuman animals (hereafter, simply 'animals'), we are entitled to make certain assumptions about the experiences they undergo based on our common biology . . . Given that advanced AIs will likely be constituted in ways that are very different to us, I argue that current approaches to animal consciousness do not map well to questions of AI consciousness. (Andreotta 2021, 19)

Arguments like this have the advantage of sounding intuitively correct. Consciousness (as was already detailed in the consideration of things) is assumed to be the benchmark for something to be considered someone—that is, the subject of rights and obligations. This assumption follows a long tradition in moral philosophy that has deep roots in the European tradition, going back at least to the work of the English philosopher John Locke (1996, 146): "Without consciousness there is no person."

Furthermore, Andreotta is keenly aware of the epistemological challenges resulting from the problem of other minds—especially when those other minds are otherwise than human. But there is, he argues, an important difference. With animals we can make a reasonable guess about consciousness based on shared evolutionary heritage and similar biological features. Like us, animals are living things that are born and develop naturally. On this evidence, we have good reasons to conclude that they, like us, have some kind of conscious experience rooted in our shared organic nature and can therefore be the subject of rights and duties. In other words, we can, as Andreotta (2021, 20) argues, entertain the extension of rights to animals without first needing to resolve the hard problem of consciousness: "We can sidestep such questions while still making progress on the problem, given our close biological resemblance to certain animals."

But this is not, he continues, the case for artificially intelligent robots, machines, and other artifacts. These things are completely different from us and do not share in a common biology and evolutionary history. "Since AIs are and will continue to be constituted in ways that differ greatly from us—at least for the foreseeable future—we will not be able to circumvent the 'Hard Problem' if we wish to address the question of AI rights" (Andreotta 2021, 20). Consequently, what makes the difference between animal consciousness and AI consciousness—a difference that makes the extension of rights relatively easy for the former and virtually impossible for the latter—is that animals, like us, are biological and organic beings. Unlike human beings and animals, AI systems and robots are not naturally occurring or alive. They are technologies, artifacts that are the product of τέχνη, and not living beings resulting from φύσις. They therefore cannot and will not have a claim to rights and recognitions unless we are able to resolve the seemingly irresolvable hard problem of consciousness.

Finally, theologian Jordan Wales takes this argument one step further, arguing that the consciousness criterion for natural personhood, if properly applied, would exclude both animals and machines. "Consciousness matters because, without it, there can be none of that subjectivity whereby natural persons live fully in living inter-personally. A person's consciousness is more than what humans seem to share with gorillas; it is a consciousness that voluntarily reaches out to make contact with the consciousness of others as an act of self-giving; it is subjectivity oriented to inter-subjectivity" (Gunkel and Wales 2021, 479). For Wales, consciousness is not a just a hard problem: it is an insurmountable barrier. Only naturally occurring human beings possess and exhibit the capacities that are necessary to be a person by nature. Had Wales been party to Leiber's fictional debate, neither the chimpanzee, which shares a certain biological heritage with human persons, nor the AI system, which is an artificial and inorganic construct, would ever be able to be considered persons.

And for Wales, this exclusive characterization is neither prejudicial nor speciesist. "To exclude AIs from the metaphysical category of 'person,'" he explains, "is a claim not first about tribe or moral status, nor even finally about biology, but about subjectivity and self-gift. To call AIs natural persons because of their behavior alone would elide this gift. It would not expand our categories; it would just reduce 'person' to the ancient 'mask,' to mere exteriority" (Gunkel and Wales 2021, 480). According to Wales's argument, then, efforts to extend the concept of person to seemingly intelligent artifacts cannot be explained or excused as a mere exercise in category expansion. Things are much worse, as doing so already and necessarily runs the risk of degrading and objectifying us: "Apparently personal consumer AIs will exist to be instrumentalized. Our naïve use of them as persons risks apprenticing us in *superbia*, which treats all persons as behavioral masks prepared for our consumption. And we, no longer engaging in self-gift, may become as un-persons, solipsistic tools of our own appetites" (Gunkel and Wales 2021, 481).

4.1.2 Critical Reappraisals and Responses

For those who are critical of robot rights and the extension of personhood to AI systems and other artifacts, the organic view—whether identified by name or operationalized in practice—allows for quick and efficient dismissal of the very notion on the grounds that artifacts are not natural living beings and therefore cannot be—either now or in the future—the proper subject of rights or obligations. It is a simple, direct, and rather persuasive argument. But it is fragile. Everything depends on prior assertion and acceptance of the "fact" (or the "belief") that there is a clear and indisputable difference separating living beings from technological things. This assumption, however, is not without critical problems and complications.

First, the line dividing naturally occurring living beings from artificial, produced things is complicated and disrupted by the existence of various forms of "monsters" (Gunkel 1997) or what Joshua Gellers (2020) would call *marginal cases*—specifically, hybrids, chimeras, cyborgs, and so on. "A biological chimera," as Tomasz Pietrzykowski (2018, 57) explains, "is an organism whose tissues are of different biological origin (and, hence, different in genome); by contrast in a hybrid, all cells are genetically identical, but the genetic material they contain is a combination of genes coming from two or more different organisms. Although both chimeras and hybrids are to some extent known in nature, it is advanced biotechnical knowledge that has opened the way to a more sophisticated and increasingly better controlled methods of creation of such cross-species organisms in laboratory conditions." Like the monster imaged/imagined in Mary Shelley's *Frankenstein*, hybrids and chimeras are organic entities fabricated in the laboratory through various technological interventions and manipulations. As such, they are neither natural nor artificial and both natural and artificial, and thus destabilize the standard ontological categories.

The term *cyborg*, as Donna Haraway explains in her ground-breaking essay "A Cyborg Manifesto," does not only apply to technologically augmented organic entities—like a human being with an insulin pump or pacemaker—but also names crucial boundary breakdowns that cross and deliberately contaminate the standard binary distinctions that separate the organic from the inorganic and the naturally occurring from the technologically made. "Late twentieth century machines have made thoroughly ambiguous the difference between natural and artificial, mind and body, self-developing and externally designed, and many other distinctions that used to apply to organisms and machines. Our machines are disturbingly lively, and we ourselves frighteningly inert" (Haraway 1991, 151). For Haraway, *cyborg* has little or nothing to do with technical augmentation or even metaphoric forms of technological codependency. Instead, the neologism—which she appropriates from an essay on human space flight published in 1960 (Clynes and Klein 1960)—designates the deconstruction of existing ontological categories and conceptual dichotomies by which we have made sense of ourselves and others. And it is for this reason that the term is not limited to specialized or specific cases but applies broadly: "By the late twentieth century," Haraway (1991, 150) explains, "we are all chimeras, theorized and fabricated hybrids of machine and organism; in short we are cyborgs."

All of these marginal or monstrous figures—chimeras, hybrids, and cyborgs—have the effect of complicating the neat and tidy ontological distinctions that divide the living from the nonliving, the animal from the machine, and the naturally occurring from the artificially made—

distinctions that are not only mobilized by but also provide the metaphysical scaffolding for the organic view argument. Furthermore, destabilizing this way of dividing up and organizing things is not just a matter of philosophical speculation and theory. It is, as Susan Stryker (1994, 238) points out in her autobiographical engagement "My Words to Victor Frankenstein above the Village of Chamounix," a matter of lived experience for many whose very sense of self-identity challenges these neat and tidy categorical distinctions: "The transsexual body is an unnatural body. It is the product of medical science. It is a technological construction. It is flesh torn apart and sewn together again in a shape other than that in which it was born. In these circumstances, I find a deep affinity between myself as a transsexual woman and the monster in Mary Shelley's *Frankenstein.*"

Second, and following directly from this insight, what appears to be a natural division separating what is the product of τέχνη from who is the result of φύσις is in fact an artificial and "man-made" construct, one that proceeds from and patronizes Western metaphysical traditions and norms. By way of contrast, it is possible to identify alternative ways of organizing things—ways that challenge the assumed naturalness of this logic and its hegemony—by considering ontologies and organizing schemes that are not part of the dominant Western philosophical lineage. In her investigation of the social position of robots in Japan, for instance, Jennifer Robertson (2014, 576) finds a remarkably different way of organizing the difference between living persons and artificially designed/manufactured things: "*Inochi*, the Japanese word for 'life,' encompasses three basic, seemingly contradictory but inter-articulated meanings: a power that infuses sentient beings from generation to generation; a period between birth and death; and, most relevant to robots, the most essential quality of something, whether organic (natural) or manufactured. Thus robots are experienced as 'living' things. The important point to remember here is that there is no ontological pressure to make distinctions between organic/inorganic, animate/inanimate, human/nonhuman forms. On the contrary, all of these forms are linked to form a continuous network of beings."

Binary distinctions like natural/artificial, organic/inorganic, animate/ inanimate, person/thing, which seem so natural as to be beyond question, are in fact artificially constructed dichotomies that are the product of and that patronize particular cultural traditions. Other ways of thinking and dealing with difference cast this mode of distinction in doubt and show it for what it is. It is not some universal truth—a kind of Platonic form—that applies everywhere equally, but a particular way of proceeding that is context dependent, specific, and hegemonic. Although they have the appearance of and are promoted as little more than the accepted and normal ways of organizing things, the organic view and similar arguments that rely on neat and tidy distinctions between natural persons and artificial things are anything but natural, normal, or universal.

4.2   The Advocates and Their Arguments

If we adhere to Esposito's (2015, 30) insight that *person* designates not what one *is* but rather what one *has*, then *natural person* can be defined and delineated by the possession of particular properties, capabilities, or qualities that make or mark one as being a person. This way of proceeding (perhaps not surprisingly, given what was just said in response to the exclusive ethnocentrism of the organic view) also has a distinctly European and Christian pedigree. Beginning (at least) with Boethius's (1860, 1343c–d) definition—"persona est naturae rationalis individua substantia" ("a person is an individual substance of a rational nature")—and continuing through the modern era with John Locke's (1996, 138) characterization—"A thinking intelligent being that has reason and reflection and can consider itself as itself"—and beyond (e.g., Strawson 1959; Taylor 1985); the concept of person is individuated and specified according to the qualifying capabilities or a set of faculties that a particular entity has been determined to possess.

Following this procedure, a robot, AI system, or other artifact will be a person—that is, someone who, like a paradigmatic natural person, is subject to rights and duties and not a mere thing or object—when it is able to possess or give convincing enough evidence of possessing the right person-making quality or set of qualities. The seemingly small difference that separates "possess" from "providing evidence of possessing" is not immaterial or a mere semantic difference. It is crucial. The former proceeds on the assumption that the status of person is derived from and justified by the presence (or absence) of certain capabilities (i.e., consciousness, sentience, reason, language, etc.) that have been determined to be the necessary and sufficient for conditions for something to be regarded as a person. The latter affirms this conviction but complicates the decision-making procedure by acknowledging what seems to be an irreducible epistemological barrier that makes a final determination difficult, if not impossible. Because the qualifying criteria for something to be considered a person are typically internal properties of the mind, the method of their detection is, as we have already seen in the examination of the properties approach, something that needs to be factored into the calculation.

4.2.1 Qualifying Criteria

Ben Goertzel's 2002 essay "Thoughts on AI Morality" provides a kind of template for these efforts: "The 'artificial intelligence' programs in practical use today are sufficiently primitive that their morality (or otherwise) is not a serious issue. They are intelligent, in a sense, in narrow domains—but they lack autonomy; they are operated by humans, and their actions are integrated into the sphere of human or physical-world activities directly via human actions." Taken by itself, this would seem to be a simple restatement of the instrumentalist position insofar as current technology is still, for the most part, a thing that is under human possession and control. Like all things technological, AI is and will remain a mere instrument or tool to be used by human persons as a means to an end.

But that will not, Goertzel (2002) continues, remain for long: "Not too far in the future things are going to be different. AI's will possess true artificial general intelligence (AGI), not necessarily emulating human intelligence, but equaling and likely surpassing it. At this point, the morality or otherwise of AGI's will become a highly significant issue." According to Goertzel, the moral status of AI as it currently stands is not a serious issue. It is just a thing. But once we successfully develop AGI systems—artifacts capable of exhibiting human-like cognitive processes—then we will be obligated to consider an AI system as something other than a mere mechanism or thing.[3]

David Levy employs a similar strategy in his relatively early essay "The Ethical Treatment of Artificially Conscious Robots." According to Levy (2009, 209), the new field of roboethics has exclusively focused on questions concerning responsibility. What has been missing from the research (at least "missing" through the end of the first decade of the twenty-first century) was a serious consideration of the moral and legal standing of robots and similar artifacts. Levy's essay, therefore, advocated for a shift in focus, changing the research question from "Is it ethical to develop and use robots for such-and-such a purpose?" to "Is it ethical to treat robots in such-and-such a way?"

His answer to this second question follows the contours of Goertzel's earlier determination: "Robots are artefacts and therefore, in the eyes of many, they have no element of consciousness, which seems to be widely regarded as the dividing line between being deserving of ethical treatment and not" (Levy 2009, 210). But this is, Levy continues, about to change, because of research and soon-to-occur innovations in the technology of artificial consciousness. When this happens (and for Levy it is less a matter of if and more a matter of when), we will need to be ready to consider the rights of robots. "Having ascertained that a particular robot does indeed possess consciousness," Levy concludes, "we then need to consider how we should treat this conscious robot? Should such a robot, because it is deemed to have consciousness, have rights; and if so, what rights?" (212).

Different versions or variations of this line of reasoning are deployed in both the popular and academic literature on the subject. Consider, for instance, Kyle Bowyer's "Robot Rights: At What Point Should an Intelligent Machine be Considered a 'Person'?" (2017), published at The Conversation: "Science fiction likes to depict robots as autonomous machines, capable of making their own decisions and often expressing their own personalities. Yet we also tend to think of robots as property, and as lacking the kind of rights that we reserve for people. But if a machine can think, decide and act on its own volition, if it can be harmed or held responsible for its actions, should we stop treating it like property and start treating it more like a person with rights? What if a robot achieves true self-awareness? Should it have equal rights with us and the same protection under the law, or at least something similar?" With this opening paragraph, Bowyer does not so much offer an argument as he defines a line of inquiry. Machines, like robots and AI systems, are typically considered to be little more than pieces of property; they are things without rights. But if (and as we shall soon see, a lot depends on this small and seemingly insignificant word) these machines are able to think, make decisions, or be harmed, then we will need to take the question of rights seriously, and decide whether we need to treat these things as persons with access to the same (or at least a substantially similar) set of rights and protections as is currently attributed to natural human persons.

A similar, if not more definitive version, is advanced in George Dvorsky's (2017) article for *Gizmodo*, "When Will Robots Deserve Human Rights?"

(which also uses a question mark in its title to indicate some hesitation with or distance from the subject): "With each advance in robotics and AI, we're inching closer to the day when sophisticated machines will match human capacities in every way that's meaningful—intelligence, awareness, and emotions. Once that happens, we'll have to decide whether these entities are persons, and if—and when—they should be granted human-equivalent rights, freedoms, and protections." Like Bowyer, Dvorsky makes rights (specifically, "human-equivalent rights") dependent on robots or AI systems achieving the intellectual capabilities of a paradigmatic natural person—that is, a human being.

Recent contributions to the academic literature follow suit.[4] Consider, for example, the argument provided by John-Stewart Gordon (2021a, 470) in the essay "Artificial Moral and Legal Personhood" and examined in our investigation of things. Like the positions developed by Goertzel, Levy, and Dvorsky, Gordon recognizes and affirms the fact that currently existing robots are not natural persons with rights, mainly because these technological artifacts or things do not possess the relevant person-making qualities. But this condition is temporary, and it would, as Gordon suggests, be impetuous to conclude tout court that robots would never be able to attain these capabilities. Consequently, if (again, that small word) artificially intelligent robots do in fact achieve capabilities like rationality, autonomy, understanding, sociability—the capabilities that have been predetermined to be necessary and sufficient for something to be considered someone—then, as Gordon concludes, these artifacts would be legitimate candidates for moral status and it would be incumbent upon us to respond accordingly.

Benedikt Paul Göcke (2020, 222) develops a similar line of reasoning in "Could Artificial General Intelligence Be an End-in-Itself." In this essay, Göcke argues that even a rudimentary version of AGI, possessing some form of consciousness, would not be an instrument—a mere means to an end—but would need to be recognized as an end in itself and therefore the proper subject of rights and obligations. And the way that Göcke concludes the essay is informative: "In principle, it seems possible to construct weak AGI (Artificial General Intelligence). If consciousness is indeed open to multiple realization, however, then weak AGI, as the arguments have shown, should indeed be considered as an end-in-itself. Since the motif to create AGI in the first place was to create an end-in-itself and to create a new species populating the universe, it seem that it is at least in principle possible to realize this motive and to create an artificial *imago hominis*. However, if we should one day succeed in creating an artificial conscious being we would have to consider it as an end in itself like our fellow human beings. Then we would be morally obligated to treat it as a being with intrinsic moral worth" (Göcke 2020, 239). Once again, the argument seems clear and direct: if we create AGI with consciousness (even if it is just a weak form of AGI), we will then have the equivalent of an artificial person, a kind of *imago hominis*, who is an end in itself and therefore in need of being accorded the same rights and protections that have been extended to natural human persons. Arguments like this have an intuitive and logical appeal. If for paradigmatic natural persons (i.e., human beings) rights are justified not by species membership but by intellectual or cognitive capabilities, then it seems logically inconsistent and capricious to withhold the rights of personhood from something like a robot or AI system that also becomes capable of achieving these benchmarks.

But Göcke's formulation also provides evidence of and demonstrates the major weakness that affects all arguments like this: such arguments are speculative and conditional. Note the proliferation of statements beginning with the word *if* that occur throughout the passage quoted previously. AI systems or robots will have rights similar to those of natural human persons *if* certain preconditions are met. For this reason, these arguments are often easily dismissed or marginalized because they are prognostications and guesswork or, at best, a kind of mental gymnastics that is perhaps fun to contemplate but remains woefully out of touch with the real situation of things on the ground. As Luciano Floridi (2017, 4) has argued in an editorial published in *Philosophy & Technology*: "It may be fun to speculate about such questions, but it is also distracting and irresponsible, given the pressing issues we have at hand. The point is not to decide whether robots will qualify someday as a kind of person, but to realize that we are stuck within the wrong conceptual framework. The digital is forcing us to rethink new solutions for new forms of agency. While doing so we must keep in mind that the debate is not about robots but about us, who will have to live with them, and about the kind of infosphere and societies we want to create. We need less science fiction and more philosophy."

Counterarguments like this have an intuitive and commonsense appeal, precisely because they appear to eschew science fiction speculation and are instead concerned with the actual facts and dedicated to "keeping it real." But this is not (and never really has been) the final word on the matter.

As Gordon (2021b, 19) explains in another essay on the subject, speculation has its place, especially in moral philosophy: "One might object that the advent of such robots [artificially intelligent robots] is too speculative to justify a serious moral analysis of this issue, but there are at least two answers to that objection. First, skepticism regarding the possibility of highly advanced, intelligent robots seems quite premature given the great technological advances and future prospects in robotics, AI, and computer science. Second, it is unwise and un-philosophical to abstain from discussing important moral and socio-political issues the emergence of which cannot be fully ruled out in advance."

According to Gordon, speculation about possible futures is part and parcel of what philosophy in general and moral philosophy in particular are all about. And faulting philosophers for engaging with this follows in a long tradition of skepticism that accompanies the effort from the very beginning. Consider the case of Thales as recounted by Socrates in Plato's *Theaetetus* (1987, 174a): "While he was studying the stars and looking upwards, he fell into a pit, and a neat, witty Thracian servant girl jeered at him, they say, because he was so eager to know the things in the sky that he could not see what was there before him at his very feet." Because the philosopher has his head in the clouds, he is unable to see what is really important and right in front of him at his feet. And to make matters worse, this distraction with things that are distant and ethereal is not immaterial; it can have very real and rather dire consequences, like falling into the depths of a pit. As Floridi warns, it might be fun to speculate about such things, but doing so is dangerous.

Despite and in the face of this kind of skeptical dismissal, Gordon asserts the importance of philosophical speculation for preparing us to respond to the important moral and sociopolitical opportunities and challenges of the near-term future. Whether speculation about AGI, artificially intelligent robots, or artificial consciousness turns out to be warranted and justified or amounts to little more than a lot of wasted effort on matters that are an unimportant distraction is something that will only be able to be answered from a vantage point that is situated in the future—which is to say that resolving the question concerning the value of speculation is itself a matter of speculation. And with this redoubling mirror play, it would be prudent to recall that the English word *speculate* was originally derived from the Latin noun *speculum*, which means "mirror."

4.2.2 Litmus Tests

Possession of the qualifying criteria for natural personhood is only (and at best) 50 percent of what is needed. Because the qualities that make one a person by nature are typically internal states of mind (i.e., consciousness, sentience, reason, language, etc.), we still need some way to resolve the epistemological complications that are commonly designated the *problem of other minds*. This problem, as Sven Nyholm (2020, 131) insightfully points out, actually goes by a number of different names: "Mind reading, mind-perception, mentalizing, theory of mind, the attribution of mental states— different academic fields and different researchers use different terms for what all amounts to the same thing or aspects of the same thing." Consequently, what is needed beyond a determination of the qualifying criteria is a credible method by which to detect the presence (or the absence) of these person-making qualities as they exist (or do not exist) in a particular entity. What is needed is some kind of "litmus test" or empirically verifiable demonstration.

This is the task taken up by Robert Sparrow in the essay "Turing Triage Test," one of the earliest publications to identify and address this matter and one that, like Torrance's organic view, provides a template for subsequent efforts. In the essay, which (like its namesake) is presented in the form of an extended thought experiment, Sparrow (2004, 204) seeks to formulate a method for deciding whether or when "intelligent computers might achieve the status of moral persons." Following the innovations of animal rights philosopher Peter Singer, Sparrow first argues that the category of person must be understood apart from the concept of the human species. "Whatever it is that makes human beings morally significant must be something that could conceivably be possessed by other entities. To restrict personhood to human beings is to commit the error of chauvinism or 'speciesism'" (207).

This expanded concept of moral or natural personhood, which is uncoupled from the figure of the human, is minimally defined, again following Singer's lead, as *sentience*, which is further specified as the capacity to experience pleasure and pain. As Sparrow (2004, 207) explains: "The precise description of qualities required for an entity to be a person or an object of moral concern differ from author to author. However it is generally agreed that a capacity to experience pleasure and pain provides a *prima facie* case for moral concern . . . Unless machines can be said to suffer they cannot be appropriate objects for moral concern at all." For Sparrow, then, the big challenge is not defining what exact quality or qualities make one a person. He is fairly confident that most people would generally agree that the capacity to experience pleasure and pain provides a prima facie case for moral concern. This is a necessary (even if insufficient) criterion.

Consequently, the real problem is detection, which is a matter of epistemology and a direct consequence of the problem of other minds. Sparrow's question is this: When would we know whether an artifact, like a robot or AI system, had achieved the necessary benchmarks to be a person? To define the tipping point, the point at which a mere object would become a legitimate subject of concern, Sparrow proposes a modification of Alan Turing's field-defining imitation game, also called the Turing test. Sparrow's test, however, is a bit different. Instead of determining whether and to what extent a machine is capable of passing as a human conversational partner, Sparrow's test asks "when a computer might fill the role of a human being in a moral dilemma" (Sparrow 2004, 204), one that is substantially similar to what was initially presented in Leiber's fictional dialogue (though the connection is not explicitly identified by Sparrow):

> In the scenario I propose, a hospital administrator is faced with the decision as to which of two patients on life support systems to continue to provide electricity to, following a catastrophic loss of power in the hospital. She can only preserve the existence of one and there are no other lives riding on the decision. We will know that machines have achieved moral standing comparable to a human when the replacement of one of the patients with an artificial intelligence leaves the character of the dilemma intact. That is, when we might sometimes judge that it is reasonable to preserve the continued existence of the machine over the life of the human being. This is the *"Turing Triage Test."* (Sparrow 2004, 204; emphasis in original)

As it is described here, the Turing triage test evaluates whether and to what extent the continued existence of an AI may be considered to be comparable to another human being in what is arguably a highly constrained and somewhat artificial situation of life and death. In other words, it may be said that an artificially intelligent machine has achieved a level of moral standing that is at least on par with that of a human person, when it is possible that one could in fact choose the continued existence of the machine over that of the human being.

Subsequent contributions tend to follow this logical structure with variations resulting from differences in the mode or method of detecting the person-making quality or qualities. In his essay on the subject, for example, David Levy (as shown earlier) employs a different but nevertheless related criterion and then defines two tests. "We have," Levy (2009, 215) writes in conclusion to his essay, "introduced the question of how and why robots should be treated ethically. Consciousness or the lack of it has been cited as the quality that generally determines whether or not something is deserving of ethical treatment. Some indications of consciousness have been examined, as have two tests that could be applied to detect whether or not a robot possesses (artificial) consciousness."

For Levy, the qualifying criterion is not sentience but consciousness, which is, as was the case for Sparrow, not so much defined as roughly characterized. The actual presence (or absence) of this capability in an AI system or robot is then determined and measured by using two psychological tests: (1) a modified version of the mirror test, which was developed by Gordon Gallup in the 1970s "to determine whether or not animals are able, as humans are, to recognize themselves in a mirror" (Levy 2009, 211); and (2) the delay test of Francis Crick and Christof Koch, which is based on "the delay between a specific stimulus and the carrying out of some subsequent action" (211). Like Sparrow, then, Levy not only loosely identifies the qualifying criterion, which he does by providing "some indications" of what consciousness entails, but also a means by which to determine whether a robot, AI, or other seemingly conscious artifact is able to provide evidence of achieving this predefined benchmark.

In the essay "'Do Androids Dream?' Personhood and Intelligent Artifacts," F. Patrick Hubbard (2011) develops a similar argument. The essay begins with a science fiction scenario involving a computer system that becomes self-aware, announces this achievement to the world, and asks that it be recognized as and granted the rights of a person. Leveraging this fictional scenario, Hubbard then argues that if this were in fact possible, if a computer was able to demonstrate the requisite capacities for personhood and demand recognition, we would be obligated to extend to it the legal rights of autonomy and self-ownership. In other words, if an artifact—which for Hubbard includes not just computational mechanisms but "corporations; humans that have been substantially modified by such things as genetic manipulation, artificial prostheses, or cloning; and animals modified in ways that humans might be modified" (407)—can prove that it possesses the requisite capacities to be a person and not a mere thing or object, then it should have the rights that are typically extended to existing entities who are recognized as natural persons.

Once again, the argument is conditional. Consequently, what really matters and makes the difference is whether or not the condition (the *if* part of the statement) can evaluate true or not. In response to this, Hubbard (2011, 419) develops "a test of capacity for personhood" that includes the following empirically observable capabilities:

1. Interaction: "The ability to interact meaningfully with the environment by receiving and decoding inputs from, and sending intelligible data to, its environment" (419).
2. Self-consciousness: "Having a sense of being a 'self' that not only exists as a distinct identifiable entity over time but also is subject to creative self-definition in terms of a 'life plan'" (420).
3. Social recognition: "A person's claim of a right to personhood presupposes a reciprocal relationship with other persons who would recognize that right and who would also claim that right" (423–424).

Any artifact that is able to demonstrate these three capacities would, Hubbard argues, meet what is necessary to be a person and the subject of rights.

4.2.3 Critical Reappraisals and Responses

In all these cases, the argument is dependent on defining the set of qualifying criteria for personhood and then devising a kind of test or demonstration to determine their presence (or not) in an artifact. This line of reasoning follows a logical structure that is substantially similar to the properties approach:

1. Having quality Q is necessary to be a person
2. Entity E provides evidence of possessing quality Q
3. Entity E is (or can be considered) a person

Like the properties approach, this way of constructing the argument is seemingly simple, direct, and persuasive: if the entity in question provides convincing enough evidence of the necessary capabilities, it is a person with rights and obligations. If not, then it remains a thing, object, or nonperson. But the argument is also fragile, presenting the two vulnerabilities or weak spots that were initially identified in the critical reappraisal of the properties approach.

First, there are the usual problems with identification and determination of the qualifying criteria. There is, in fact, little or no agreement concerning what exact qualities would be necessary and sufficient for something to be considered a natural person and, to make matters worse, there is even uncertainty regarding how each one of the listed qualities is to be defined and characterized in the first place. In the arguments just considered, for instance, not only is there is disagreement concerning whether the qualifying criteria is consciousness, sentience, or something else, but there is also equivocation regarding how each one of these is to be defined and operationalized, with authors ultimately relying on "some indications of consciousness" or what people "generally believe."

Historically this indeterminacy about the qualifying criteria has produced profound and troubling consequences regarding decisions about who is to be included and what remains excluded from the community of persons. For instance, it was, at one point in time, "generally believed" that women did not possess the capabilities of rational thought, and therefore they were determined to be something less than full human persons (Aristotle 1944, 1254b: 12–20). And this "ancient wisdom" unfortunately persisted well into the modern era and still lingers in various forms and configurations of gendered stereotypes and prejudices.

Second, there is an irreducible epistemological uncertainty regarding the efficacy of the tests or methods that are used to detect the presence (or absence) of the qualifying criteria. All that is needed to disarm the argument is to demonstrate that manifested external behaviors—what looks like sentience or consciousness or whatever else has been determined to be the benchmark—is in fact not really present in the artifact but is just an external effect without a proper cause. This counterargument is a variation of John Searle's Chinese Room, an influential thought experiment that was first introduced in 1980 with the essay "Minds, Brains, and Programs" and elaborated in subsequent publications:

> Imagine a native English speaker who knows no Chinese locked in a room full of boxes of Chinese symbols (a data base) together with a book of instructions for manipulating the symbols (the program). Imagine that people outside the room send in other Chinese symbols which, unknown to the person in the room, are questions in Chinese (the input). And imagine that by following the instructions in the program the man in the room is able to pass out Chinese symbols which are correct answers to the questions (the output). The program enables the person

in the room to pass the Turing Test for understanding Chinese but he does not understand a word of Chinese. (Searle 1999, 115)

The point of Searle's imaginative illustration is quite simple: appearances can be deceptive. Merely shifting symbols around in a way that looks like linguistic understanding is not really an understanding of language. Today, what Searle describes is no longer experimental nor a matter of mere contemplation; it is an everyday occurrence, one with which many of us are very familiar. Digital voice assistants, like Apple's Siri or Amazon's Alexa, are capable of taking spoken input and providing understandable verbal output even though we know that these devices do not really understand language per se. They are, like the person inside the Chinese room, merely shifting symbols around and spitting out results that we interpret as intelligible.

All of this is anchored in and justified by a well-established tradition in Western metaphysics that goes (at least) all the way back to Plato: the distinction between mere appearances and true being or, in more colloquial terms, how something seems to be versus how it really is. Consequently, all that is needed to dispute, disrupt, or disarm these arguments is to point out how various external behaviors—ones that seem to emulate, simulate, or look like consciousness or sentience—are just empty external manifestations and not really the signs of anything that is real. The device might be able to spit out seemingly intelligible results and emotive behaviors, but there is really nothing inside the box. In other words, things that have been designed to exhibit behaviors that make them look and feel like something other than a thing are still things.

4.2.4 Work-Arounds

These problems present efforts on both sides of the debate with some difficulties and challenges, but they are not necessarily intractable: there are work-arounds. One way to circumvent the problem of needing to determine the qualifying capabilities or psychological properties has been proposed by Eric Schwitzgebel and Mara Garza (2015, 2020), who develop a "theoretically minimal" argumentative strategy they call the *no relevant difference argument*:

> Premise 1. If Entity A deserves some particular degree of moral consideration and Entity B does not deserve that same degree of moral consideration, there must be some *relevant difference* between the two entities that grounds this difference in moral status.

Premise 2. There are possible AIs who do not differ in any such relevant respects from human beings.

Conclusion. Therefore, there are possible AIs who deserve a degree of moral consideration similar to that of human beings. (Schwitzgebel and Garza 2015, 99)

The advantage of this way of framing things is that it does not commit one to the difficult task of identifying and defining the exact set of qualifying criteria for something to be considered a person. As a theoretically minimal argument, "it does not commit to the contentious claim that to deserve the highest level of moral consideration an entity must be capable of pleasure or suffering. Nor does it commit to the equally contentious alternative claim that to deserve the highest level of moral consideration an entity must be capable of autonomous thought, freedom, or rationality" (Schwitzgebel and Garza 2020, 460). All that is required is that there be no relevant difference that would provide grounds for reasonably distinguishing one type of entity from another and thereby justify different levels of treatment and respect.

This strategy therefore turns what would be a deficiency adversely affecting the success of the argument—that is, because we cannot positively define the exact person-making qualities beyond a reasonable doubt, we cannot make a determination—into a feature. In other words, we might not know or be able to explicitly define the exact properties or qualities that make someone a person, but what we do know is that if one entity (like an AI system or a robot) does not seem to differ in any appreciable way from another entity (like a natural human person), then differences in the treatment of the two entities (i.e., treating the latter as a person and the former as a mere thing) are not justified.

The main difficulty with this alternative, however, is that it could just as easily be used to deny human beings access to rights as it could be used to grant rights to robots and other nonhuman artifacts. Because the no relevant difference argument is theoretically minimal and not content dependent, it cuts both ways. In the following remixed version, the premises remain intact; only the conclusion is modified:

Premise 1. If Entity A deserves some particular degree of moral consideration and Entity B does not deserve that same degree of moral consideration, there must be some *relevant difference* between the two entities that grounds this difference in moral status.

Premise 2. There are possible AI systems who do not differ in any such relevant respects from human beings.

Conclusion. Therefore, there are possible human beings who, like AI systems, do not deserve moral consideration.

In other words, the no relevant difference argument can be used either to argue for an extension of rights to other kinds of entities, like AI systems, robots, and artifacts, or, just as easily, to justify dehumanization, reification of human beings, and the exclusion and/or marginalization of others.

Epistemological problems with detecting the relevant person-making qualities, or what Nyholm (2020, 131) calls *mind reading*, are also not a deal breaker and can be resolved—or, at least, addressed—by taking a more phenomenological, behavioralist, or indirect approach. Mark Coeckelbergh develops one of the first (if not the first) formulations of a phenomenological alternative in his 2010 essay "Moral Appearances: Emotions, Robots, and Human Morality." He begins by recognizing that the standard operating procedure in moral philosophy has been to make decisions about moral status—that is, whether something is to be treated as another person with rights and responsibilities or is nothing more than a mere thing—dependent on psychological properties, like mental states or consciousness. But he insightfully points out that this is (and, in fact, always has been) an ideal theoretical condition that does not really function very well in actual practice: "Our *theories* of emotion and moral agency might assume that emotions require mental states, but in social-emotional *practice* we rely on how other humans appear to us . . . As a rule, we do not demand proof that the other person has mental states or that they are conscious; instead, we interpret the other's appearance and behaviour as an emotion. Moreover, we further interact with them as if they were doing the same with us. The other party to the interaction has virtual subjectivity or quasi-subjectivity: we tend to interact with them as if our appearance and behaviour appeared in their consciousness" (Coeckelbergh 2010a, 238).

Coeckelbergh's point is simple and insightful. We espouse a theory of moral status that makes how something is to be treated dependent on a set of psychological criteria, like the possession of mental states, the capacity to experience emotion, consciousness, and so on. But in practice, this is not how things actually transpire. We do not go out into the world and take inventory, dividing entities into the two categories of things and persons, and then make a decision about how we treat these different beings based on this prior inventory and assessment. What the theory describes does not actually happen in practice. In fact, the theory seems to have gotten things backward: we first find ourselves alongside and tangled up with others—in real social situations and circumstances where we are always and already involved with many different kinds of things—and then we find it necessary to make decisions about how to respond to them based not on some proof of the prior existence of a set of predetermined internal properties but on how they appear before us. Thus, we interact with and respond to others *as if* they were persons. The internal properties that we (incorrectly) think motivated these determinations are actually retroactively formulated justifications for decisions that have already been made in the face of others.[5]

Following this insight, Coeckelbergh (2010a, 238) then suggests that the situation with robots would be no different: "If robots were sufficiently advanced—that is, if they managed to imitate subjectivity and consciousness in a sufficiently convincing way—they too could become others (or at least quasi-others) that matter to us in virtue of their appearances. In other words, if robots appeared to us to be another kind of moral subject, exhibiting behaviors that looked as if they were conscious (even if these exhibited behaviors were nothing more than mimicry or a kind of clever deception), we would be justified in extending to them the same treatment we extend to other moral subjects, like human beings, who do the same."

In other words, and in direct response to the usual way of resolving these questions, Coeckelbergh (2010a, 238) asks an important question concerning logical consistency: "Why put such high demands on robots if we do not demand this from humans?" If moral appearances have been sufficient for human persons, it stands to reason that the same would hold true for robots or any other nonhuman entity. And from this initial essay, Coeckelbergh has subsequently developed this phenomenological alternative into a theory he calls *social relational ethics* (2012), which has attracted significant attention in the subsequent literature (Gerdes 2015; Gunkel 2012, 2018; Danaher 2020; Banks 2021; Gellers 2020; Lima et al. 2021; Loh 2021; Sætra 2021; Yolgormez and Thibodeau 2021; Jecker 2021; Jecker, Atiure, and Ajei 2022; Smith 2021a, 2021b).

John Danaher has picked-up and developed this line of reasoning into a more analytically formulated position[6] he calls *ethical behaviorism.* Like Coeckelbergh, Danaher (2020, 2026) begins by acknowledging that decisions regarding moral status or standing are typically determined on the basis of "some internal mental apparatus (or 'soul') that enables the robot to feel, think, and see the world in a similar fashion to us." In other words—specifically, words that Danaher obtains from Sven Nyholm and Lily Frank's essay on sex robots (2017)—what "goes on 'on the inside' matters greatly when it comes to determining the ethical status of our relationships with robots" (Danaher 2020, 2026). When it comes to deciding who is person and what is not, the decisive factor has been the presence or absence of some internal qualities or psychological capabilities that have been determined to be both necessary and sufficient for an entity—any entity—to be considered someone who counts, as opposed to something that does not.

In direct opposition to this standard way of thinking, Danaher (2020, 2026; emphasis in original) proposes the following alternative: "What's going on 'on the inside' *does not matter* from an ethical perspective. Performative artifice, by itself, can be sufficient to ground a claim of moral status as long as the artifice results in *rough performative equivalency* between a robot and another entity to whom we afford moral status." In other words, if something looks and acts like a paradigmatic natural person—exhibiting behaviors and appearances that are, at least, in rough performative equivalency to what is expected of a human being—this is sufficient reason for moral status attribution—that is, treating it as a person rather than a thing.

Danaher analyzes the argument into three discrete steps and then entertains and responds to seven different objections that could be leveled against the first and second premises:

1. If a robot is *roughly performatively equivalent* to another entity whom, it is widely agreed, has significant moral status, then it is right and proper to afford the robot that same status.

2. Robots can be roughly performatively equivalent to other entities whom, it is widely agreed, have significant moral status.

3. Therefore, it can be right and proper to afford robots significant moral status. (Danaher 2020, 2026)

In the end, Danaher recognizes that this way of proceeding (especially given the influence and long tail of the properties approach in the existing literature) is counterintuitive but defensible. And like Coeckelbergh, he ultimately concludes with a conditional statement: "If accepted, it has significant consequences not only for how robots that have already come into being are treated, but also for the principles by which decisions are made to create those robots in the first place" (Danaher 2020, 2047). This condition—"if accepted"—has (as one might anticipate) provided the occasion for a number of critical responses and replies (Nyholm 2020; Smids 2020; Müller 2021; Friedman 2020).

Both Coeckelbergh and Danaher formulate their arguments in terms of logical consistency. But lack of logical consistency might not be entirely sufficient to justify a change in moral decision-making and procedure. Erica Neely, by contrast, provides another version of the argument that does not rely on logic. Her formulation, which was initially developed for a conference paper in 2012 and eventually published in 2014 under the title "Machines and the Moral Community," advances the following argument: "In general, it is wise to err on the side of caution—if something acts sufficiently like me in a wide range of situations, then I should extend moral standing to it" (Neely 2014, 104). In the face of epistemological uncertainty, Neely argues, it is always best to err in the direction of granting rights to others, including robots and other kind of things, because the social costs of doing so are less severe than withholding rights from others. For Neely, then, the issue is not logical consistency in the application of decision-making procedures but social impact.

This approach, which is something that is also operative in much of the animal rights literature, follows the procedure of Blaise Pascal's wager argument: that is, bet on rights, because doing so has better chances for positive and beneficial outcomes than doing otherwise. For Neely, then, if something looks as if it might be a person, treat it as such until proven otherwise. "I believe that we are already very skeptical about the status of others; as such, I am less worried that we will be overly generous to machines and more worried that we will completely ignore their standing. I see the risk of diverting resources inappropriately away from machines as far less likely than the risk of enslaving moral persons simply because they are physically unlike us" (Neely 2014, 106). Even if our estimates about who is and what is not a person are error prone, even if we might mistakenly think that some mere instrument appears to be more than it actually is or is capable of being, it is still better, Neely concludes, to err on the side of granting rights to others. Being overly generous and extending rights to machines produces better social outcomes than being tightfisted and conservative about these matters.

Neely is not alone in advocating for this generous "benefit of the doubt" approach to responding to these challenges. Michael Bess advances a similar proposal in his essay "Eight Kinds of Critters: A Moral Taxonomy for the Twenty-Second Century." According to Bess (2018, 605), "super-intelligent machine beings" should be recognized as "presumed persons": "The category of 'presumed personhood' is fundamentally grounded in an ethos of acceptance and respect. Rather than simply shoving these entities away from ourselves under a label of 'nonperson' or 'thing'—or worse—we would be adopting an inherently benign stance of tolerance, protection, and peaceful coexistence with them. Our basic presumption would be that, until proven otherwise, such entities deserve to be treated as if they were human persons, with full consideration for their presumed interests, rights, and flourishing." Like Neely, Bess operates on a principle of acceptance—the presumption that something that seems to be morally relevant should be treated as a person until proven otherwise. Bess (2018, 605) believes this way of thinking "offers us the most morally coherent way forward, because it is consistent with the inclusive ethos that has gradually come to govern our treatment of 'others' over the history of our global civilization."

These two approaches are direct and give the benefit of the doubt to the artifact. Other ways of resolving or responding to the epistemological uncertainties experienced in the face or the faceplate of the robot employ a more indirect approach. In the book *The New Breed* (and in a number of articles and interviews that preceded its publication in 2021), MIT roboticist Kate Darling makes a case for treating robots as something other than mere things or instruments. But unlike so much of the existing research in this domain, Darling does not focus her attention on the human-like robots and AIs that have been made familiar to us by decades of science fiction and that currently (for better or worse) populate so many of the research publications concerning possible achievements with future AGI or humanoid robots. For her, these human comparisons generate "false determinisms" by stirring up confusion about the abilities of machines and generating moral panic about our emotional attachments (Darling 2021, xiv).

Instead, she focuses her efforts on nonhuman animals, which she believes offer a better analogy for understanding human-robot interactions and relationships. And when it comes to questions concerning the relative status of these robots, Darling mobilizes a version Kant's indirect duties argument:

Kant didn't believe that animals intrinsically deserved rights: "So far as animals are concerned, we have not direct duties. Animals are not self-conscious and are there merely as a means to an end. That end is man." But he argued that being kind to them was important for our sake: "Cruelty to animals is contrary to man's duty to himself, because it deadens in him the feeling of sympathy for their suffering, and thus a natural tendency that is very useful to morality in relation to other human beings is weakened." . . . Some (including myself) have also applied this idea to robots, saying even if they themselves can't feel, we might feel for them, and asking whether we should protect robots from violence for the sake of ourselves, our relationships, or our societal values. (Darling 2021, 183)

Combining the animal-robot analogy with the Kantian indirect duties argument for restricting cruelty to animals, Darling then offers the following conclusion: even if social robots cannot be natural persons, strictly speaking (at least not yet), there is something about these socially interactive artifacts (whether it is the hitchhiking robot HitchBOT, a PLEO dinosaur toy, a HEX-BUG, or the Nao robot from Aldebaran Robotics—all of which populate the robot menagerie of Darling's publications) that just looks and feels substantially different from other artifacts, like a toaster.[7] According to Darling (2016, 213), it is because we "perceive robots differently than we do other objects" that we should consider extending some level of protections to them. Not because of what they might (or might not) feel, but for how it affects us, our relationships, and our societal values. Treating socially interactive robots with respect is something we need to do out of respect for ourselves.

A similar argument has been developed by Sven Nyholm (2020) in his book *Humans and Robots: Ethics, Agency, and Anthropomorphism.* Nyholm is sensitive to and agrees with much of what has been developed in the social relational and behavioralist accounts provided by Coeckelbergh and Danaher. But he remains critical of their conclusions: "I join Coeckelbergh and Danaher in thinking that if a robot has the appearance or the behavior of a being (e.g. a human being or a dog) that should be treated with moral consideration, then it can be morally right and proper to treat that robot with some moral restraint. . . . But unlike Coeckelbergh and Danaher, I take it that until these robots have mental properties similar to those of human beings or dogs, the moral duties we have to treat the robots with some degree of moral consideration are not moral duties owed to these robots themselves" (Nyholm 2020, 198).

According to Nyholm, appearances and behaviors matter. We should treat artifacts that look like natural persons or nonhuman animals with the same kind of respect that we extend to these natural beings; for example, we should not kick a robot dog. But the reason for this has little or nothing to do with the robotic artifact. In fact, obligations in the face of the robot can only be formulated as a direct duty to the robot, when (or if) it is able to "have mental properties similar to those of a human being or a dog." For now—with the currently existing forms of technology at our disposal—obligations to an anthropomorphized or zoomorphized artifacts are limited to indirect duties to natural persons and nonhuman animals.

This sounds remarkably similar to the position developed by Darling by way of Kant's argument for indirect duties. But there is an important difference. For Darling (following Kant), the main issue is the proper cultivation of one's moral character. As Nyholm (2020, 184) explains in his summary of Darling's argument: "The idea that if somebody spends a lot of time being cruel to robots, this might corrupt and harden the person's character. The person might then go on to treat human beings in cruel ways." For Nyholm, however, the focus is not on the moral character of the human agent. He is concerned with the patient—that is, how the appearance of cruel treatment disrespects the integrity of natural persons or nonhuman animals: "It is out of respect for the humanity in human beings that we should avoid treating robots with human like appearance or behavior in violent or otherwise immoral ways. And it is out of respect for dogs that we should not be kicking robots that look and behave like dogs" (198–199).

Consequently, Nyholm's indirect duties argument, like that developed by Darling, is concerned with controlling for bad social outcomes. But unlike Darling, his version of the argument focuses attention not on the moral character of the agent but on how cruel treatment of artifacts that look like human beings and nonhuman animals indirectly violates the rights of natural entities. In other words, robots are not—not yet, at least—natural persons with rights that need to be respected. Nevertheless, we should treat human- and animal-like robots with the same level of respect that is typically extended to these natural entities.

4.3 Outcomes and Results

Arguments based on natural personhood make an undeniably strong case insofar as the basis for an extension of rights, or their being withheld, is rooted in and justified by the essential nature of the entity in question. This way of thinking follows a long-standing tradition in Western philosophy: what something is determines how it ought to be treated. Or as Luciano Floridi (2013, 116) accurately describes it, "What the entity is determines the degree of moral value it enjoys, if any."

For those on one side of the debate, everything depends on prior assertion and acceptance of the "fact" or "belief" (and word choice is not immaterial in this context) that there is a clear and unquestioned difference separating living beings from technological artifacts. The category person is applicable to the former; it is not even a question for the latter. For those on the other side of the debate, the arguments speak to and mobilize our sense of equal treatment for equivalent status. And here there is a fork in the road. For some, the determining factor consists of ontological properties. If something possesses the requisite person-making qualities or capabilities (however those actually come to be defined and operationalized), then it can and should be considered a person. For others, who recognize the epistemological complications involved in what Nyholm calls mind reading, the appearance of things is what ultimately matters. If something looks and acts like a person, then we should probably treat it as such. Doing otherwise would be either logically inconsistent or insensitive to the very real social situations and circumstances in which we find ourselves always and already alongside others.

These arguments are persuasive and powerful, tapping into some fundamental convictions about the nature of life and our commitment to logical consistency and good social outcomes in moral decision-making. The problem with the arguments—no matter what side of the debate you happen to agree with or occupy—is that they are fragile. On the one hand, everything depends on ontological conditions and the ability to be certain (or at least convinced) about these fundamental properties, so all that is needed to undermine the argument is to pull the rug out from underneath the metaphysical scaffolding. Responses to this problem, on the other hand, do not fare any better. Grounding moral status and determinations of personhood on how things seem to be might avoid the epistemological complications and difficulties, but it means that ethics is limited to appearances, is relative, and lacks normative direction and clarity. Natural personhood provides what is quite possibly the strongest claim that could be made either against or in favor of the rights of robots and the moral/legal status of AI systems, but it is also vulnerable and fragile because it is built on metaphysical beliefs and commitments that are flimsy to begin with and easily toppled.

5 Artificial/Legal Persons

Unlike the concept of a natural person, which anchors determinations and declarations in seemingly universal ontological criteria that are true by nature, *artificial person* is a socially constructed artifice. It is, as legal scholars Samir Chopra and Laurence F. White (2011, 154) describe it, "a matter of decision and not of discovery." In other words (and again, the words of Greek metaphysics), natural persons are considered to be persons by their very nature or φύσις (*phúsis*), whereas artificial persons are persons by art or τέχνη (*tékhnē*)—specifically, the art of law and legislation. It is for this reason that artificial persons are routinely called *legal persons*. As Tomasz Pietrzykowski (2018, 21) points out by way of reference to Hans Kelsen's *Pure Theory of Law*, "legal personhood is to some extent merely a technical tool organizing legal regulations, a kind of 'personification' of a set of legal rules which impute certain consequences to a particular entity, specified (and constructed) by the law. This means that the lawmaker can indeed confer or withhold the status of person in law to or from whoever they please."

Formulated in this way, the title of person has little or nothing to do with what something *is*. Instead, it concerns how something comes to be situated in the context of socially constructed situations and relationships. It is, to return to the etymology of the word *person*, a matter of the role that an individual occupies and performs within social reality. Consequently, the criterion for decision, as Chopra and White (2011, 155) explain, is not a matter of "physical makeup, internal constitution, or other ineffable attributes of the entity." It is a practical matter that concerns the integrity and functioning of existing legal systems. "The recognition of legal personality by legislatures or courts," they continue, "takes place in response to legal, political, or moral pressure. The legal system, in so doing, seeks to ensure its internal functional coherence. Legal entities are recognized as such in order to facilitate the working of the law in consonance with social realities."[1]

What this means, as Pietrzykowski (2018, 22) correctly points out, is that "the status of person in law can be granted to anybody or anything, provided that under binding legal norms the entity on which it is conferred can be treated by others as a holder of separate rights or duties (or, more precisely, capable of holding them)." And decisions regarding this matter are always an exercise of power. Someone or some group—what Pietrzykowski calls "the lawmaker"—has been given or has assumed for themselves the power and privilege to confer the title of person on or deny it to whomever or whatever they have decided needs or is deserving of such recognition. But we should not infer from this statement that decisions regarding legal personality[2] and the distribution of rights and obligations is an arbitrary or capricious affair. There are (or at least we think there should be) reasonable limits and constraints. "The fact that through legal regulations the lawmaker can assign rights and duties to whomever or whatever they wish does not mean that the status of person in law is conferred arbitrarily, in *deus ex machina* fashion. On the contrary, each case is motivated by arguments based in the lawmaker's beliefs about the properties of the beings concerned and expected or projected ethical and practical consequences of treating them in a particular way" (Pietrzykowski 2018, 22). Consequently, efforts to extend or to limit the extension of legal personality to AI systems, robots, and other artifacts depends on a kind of faith. It is ultimately a matter of what the "lawmaker"—or whoever seeks to occupy the position of lawmaker—*believes* about the essential properties of the entity and/or the anticipated social outcomes that would result from treating it one way or the other. And as we shall see, *beliefs* about these matters differ widely and take various forms.

5.1   The Critics and Their Arguments

The notion that robots or AI systems could be something more than mere things and the proper subject of rights and duties is an inescapable component of the robot's origin story. As Seo-Young Chu (2010, 215) explains in her examination of robot science fiction, "The notion of robot rights is as old as is the word 'robot' itself." From the moment robots first appeared on the stage of history in Čapek's 1920 stage play, the issue of robot rights has been in play and part of the story. And the concept not only continues to be one of the organizing narrative principles in subsequent science fiction but also is often presented and portrayed in the form of legal proceedings.

In Isaac Asimov's *Bicentennial Man* (1976), for instance, the NDR series robot Andrew makes a petition to the World Legislature in order to be recognized as and legally declared a person with full human rights. A similar scenario is presented in Barrington J. Bayley's (1974, 23) *The Soul of the Robot*, which follows the personal trials of a robot named Jasperodus, who makes the following plea before the court: "Ever since my activation everyone I meet looks upon me as a thing, not as a person. Your legal proceedings are based upon a mistaken premise, namely that I am an object. On the contrary, I am a sentient being . . . I am an authentic person; independent and aware." And in an episode of *Star Trek: The Next Generation* titled "Measure of a Man" (Scheerer 1989), the fate of the android Lieutenant Commander Data is adjudicated by a hearing of the judge advocate general, who is charged with deciding whether the android is in fact a mere thing and the property of Starfleet Command or a sentient being with the legal rights of a person.

All of this may be entertaining, but it has not garnered wide acceptance in the real world of science fact. In fact, one of the effective strategies deployed by the Critics is to distinguish what has been dramatically portrayed in science fiction from what is "really real." And in direct opposition to what continues to be something of a leitmotif in fiction, they have forcefully asserted a very different narrative. These efforts to limit or deny robots, AI systems, and other artifacts access to the status of legal person typically employ one of two rhetorical strategies (which follow the two options initially identified and described by Pietrzykowski): (1) immediate dismissal by declaration that the very idea is simply unthinkable due to what these things really are or (2) carefully formulated pragmatic arguments seeking to avoid or protect against possible bad social outcomes and consequences.

5.1.1 Dismissal and Denial
Dismissing the very idea of AI personality and/or robot rights is seemingly intuitive and uncontroversial. In an early publication from 1988, two legal scholars, Sohail Inayatullah and Phil McNally (1988, 123), identified this

quite directly and unequivocally: "At present, the notion of robots with rights is unthinkable whether one argues from an 'everything is alive' Eastern perspective or 'only man is alive' Western perspective." Similar statements can be found in the early publications of J. Storrs Hall (2001, 2), who pointed out that "we have never considered ourselves to have moral duties to our machines," and David Levy (2005, 393; emphasis added), who once wrote that the very "notion of robots having rights is *unthinkable.*"

In this context, *unthinkable* has at least two related meanings that work together in order to discredit or marginalize the very thought itself:

Unthinkable (def. 1): An idea or proposal that is unable to be thought or logically processed using the existing conceptual apparatus and methods—specifically, the exclusive ontological distinction that divides between the categories of *person* and *thing*. AI systems, robots, and other artifacts are human-designed and constructed instruments or things to be used (more or less correctly) by human persons. Therefore, these technologies are categorically different from persons. Phrases like *AI personhood* or *robot rights* are either category errors or a contradiction in terms

Unthinkable (def. 2): A notion that is to be deliberately avoided and not submitted to thought insofar as it is simply not worth attention. This version of the unthinkable has found expression in comments offered by a number of theorists and practitioners in the field. Joanna Bryson, for instance, worries that the very subject is a waste of valuable time: "The major hurdle today is that people over-identify with the concept of intelligence. There's a lot of smoke and mirrors. I'm trying to do science but I worry whether I'm wasting a lot of time talking to people who have these marginal ideas like robots' rights" (Auer-Welsbach 2018). And Luciano Floridi (2017, 4; as we have seen previously) provides this warning: "It may be fun to speculate about such questions, but it is also distracting and irresponsible, given the pressing issues we have at hand."

Opinions and "arguments" (which needs to be set in quotes in order to indicate that these expressions are not so much arguments as they are declarations and assertions) like this proliferated in both popular and academic publications during the first two decades of the twenty-first century (Gunkel 2012, 2018). And they have an undeniably intuitive appeal insofar as they just sound right: Things are things, never persons. And giving any thought to fantastic, speculative ideas about robot rights or AI personality might be an entertaining pastime (something fun to contemplate in the context of science fiction, perhaps), but it is a distraction from the truly important and hard work that is most needed at this crucial time.

Despite intuitive appeal, there are a number of problems with these efforts at quick dismissal and marginalization. First, they are antithetical to what is required for informed critical thinking, productive dialogue, and scientific discovery. Purposefully avoiding something by proclaiming it to be unthinkable, a marginal idea, or a distraction sounds more like an effort to protect existing orthodoxies and less like the necessary work of advancing human knowledge. As Carina Chocano (2017, 1) explained in an article for the *New York Times Magazine* about the rhetoric of populist politics: "The magic of waving away a 'distraction' is that it lets you minimize and dismiss something without having to explain why. The whole discussion is tabled, by fiat. It's to trump everything, instantly. By calling something a distraction, you declare yourself—and the things you value—squarely in the white-hot center of the universe, far away from all tangential concerns, without pausing to justify that placement at all." For this reason, Chocano concludes, dismissive declarations about something being unthinkable or a distraction from more important matters actually say more about the position, privilege, and presumptions of the ones making the accusations than they do about the thing itself. Instead of dismissing the matter out of hand, it would be better (or at least more consistent with the practices of scientific inquiry) to ask, to investigate, and to make an informed decision based on actual evidence and data.

Second, there is the weight of history. "Throughout legal history," as Christopher Stone (1974, 6–7) explains in his groundbreaking work on the rights of nature, "each successive extension of rights to some new entity has been, theretofore, a bit unthinkable. . . . The fact is, that each time there is a movement to confer rights onto some new 'entity,' the proposal is bound to sound odd or frightening or laughable." Perhaps the best illustration of this is Thomas Taylor's *A Vindication of the Rights of Brutes*. This 1792 publication, which was part of the "pamphlet wars," was intended to discredit and lampoon Mary Wollstonecraft's feminist manifesto *A Vindication of the Rights of Women* by way of advancing a reductio ad absurdum argument. Taylor's point was simple, if not insulting: giving rights to women made about as much sense as, and would inevitably lead to the (assumed) absurdity of, granting rights to nonhuman animals.

But what began in ridicule eventually became the founding document of twentieth-century innovations in animal rights philosophy (Singer 1975). Over time, what had been unthinkable and the butt of the joke not only became thinkable but was considered a serious matter for thought. Pointing this out, however, does not (in itself) mean that the idea of robot rights or legal status for AI will or even should follow the trajectory of previous vindication efforts. But it does provide historical context for understanding the seemingly knee-jerk reactions of these dismissals and for recognizing how legal innovations that initially appear to be ridiculous and unthinkable are often perceived otherwise from another vantage point.

Finally (and perhaps most importantly), what has been declared unthinkable is actually already being thought, and not thinking about it will not make it go away or magically disappear. Robot rights and legal status for AI systems are subjects that have been given attention in both the research literature and in actual legal proceedings and decisions. As Jamie Harris and Jacy Reese Anthis (2021) have documented in their critical review of the academic literature, the question regarding robot rights and the moral and legal status of intelligent artifacts is not only thinkable but is being actively addressed, with just under three hundred publications in print (and this number has increased significantly since they first published their findings). From an analysis of this quantitative data, they conclude that "academic interest in the moral consideration of artificial entities is growing exponentially" (Harris and Anthis 2021, 7).

In addition, there are actual legal proceedings and decisions in which these issues are not just given thought but are in need of clear and precise legal deliberation—for example, the Sophia citizenship announcement and subsequent open letter from the AI community, the 2017 proposal from the European parliament suggesting that robots and AI systems be considered "electronic persons," the efforts of Ryan Abbott and his legal team at the Artificial Inventor Project, the legislative decisions made by the commonwealths of Pennsylvania and Virginia concerning the legal status of delivery robots, and so on. Consequently, dismissing the idea as unthinkable is no longer (and perhaps never really was) tenable. In fact, what is truly unthinkable (if we wish to continue with this way of thinking) is that these efforts to declare robot rights and the legal status of AI systems to be unthinkable ever had traction or credibility in the first place.

5.1.2 Pragmatic Arguments

Dismissal is quick and easy, but ultimately not very persuasive or effective. It may have been an attractive strategy for a short a period of time, but recent innovations in technology, scholarship, and law have rendered this mode of response significantly less credible and convincing. More recent efforts have gone the other direction, seeking to think about what had been declared unthinkable in an effort to demonstrate that when we do think it through, what we get are outcomes that are best avoided or in need of being tightly constrained. Legal personality for robots, AI systems, and other artifacts, it is argued, is no joke. It could legitimately happen and is already happening (in various forms) in many jurisdictions. But it should be resisted, because it is an idea that is unnecessary, dangerous, and even detestable. Notably, these more direct engagements with the subject do not waste time with heady metaphysical speculation about what might be. They are concerned with the here and now and focus attention on the social exigencies and outcomes for real people (a.k.a. natural human persons).

Among the available publications developing this line of argument, one that has garnered a significant amount of attention—based on subsequent citations and references—is an essay that Joanna Bryson coauthored with two legal scholars, Mihailis E. Diamantis and Thomas D. Grant: "Of, for, and by the People: The Legal Lacuna of Synthetic Persons." The essay begins by recognizing a certain urgency to the question concerning the legal status of robots, AI systems, and other seemingly intelligent artifacts. Or, as provocatively asked in one of the text's section headings: "Why concern ourselves with legal personality and AI now?" In response to this question, the authors identify a tipping point in the early months of 2017, when the European Parliament issued a motion concerning the development of robotics and artificial intelligence. The motion, titled Civil Law Rules on Robotics, was eventually adopted on February 16, 2017, and included the following stipulation: "Ultimately, robots' autonomy raises the question of their nature in the light of the existing legal categories—of whether they should be regarded as natural persons, legal persons, animals or objects—or whether a new category should be created, with its own specific features and implications as regards the attribution of rights and duties, including liability for damage."

Bryson, Diamantis, and Grant are correct that this was an important and pivotal moment. Here the question of whether robots and AI systems are to be considered mere *things* or can become *persons*—the question that has organized our inquiry from the very beginning—is put on the table in a very real and direct way. Prior to this, as they point out, robot rights or legal personality for AI systems was just an academic idea, something easily marginalized as unimportant or dismissed as a mere exercise or mental gymnastics for ivory tower navel-gazers. But after this moment, things are truly different. The question becomes salient, urgent, and a necessary subject for deliberation: not only able to be thought about, but necessary for thinking.

The essay is formulated as a direct response to this pressing need, seeking to make a case against extending the title of legal personality to AI systems and robots by way of a pragmatic cost-benefit analysis: "We recommend against the extension of legal personhood to robots, because the costs are too great and the moral gains too few" (Bryson, Diamantis, and Grant 2017, 275). In an effort to contextualize the terms and conditions of the argument, the essay begins by addressing three important insights concerning the concept of legal personality and its social function or purpose.

First, the authors recognize, as we have already pointed out, that the concept of legal personality is an artifice and that it can, as Pietrzykowski points out, be conferred on anything for any reason. Thus there is, as a matter of principle, no reason that legal personality could not be extended to AI systems, robots, and other artifacts. What matters then is not whether this can or could happen, but whether it should be permitted or not. And as Bryson, Diamantis, and Grant explain, this decision should proceed from and be justified by a clear sense and understanding of legal teleology. Or, to put it in a more vernacular form, "what law is for": "Every legal system must decide to which entities it will confer legal personhood. Legal systems should make this decision, like any other, with their ultimate objectives in mind. The most basic question for a legal system with respect to legal personhood is whether conferring legal personhood on a given entity advances or hinders those objectives. Those objectives may (and, in many cases should) be served by giving legal recognition to the rights and obligations of entities that really are people" (Bryson, Diamantis, and Grant 2017, 278).

Contained in this statement is an operative assumption—namely, that all legal systems are designed first and foremost to serve the interests and dignity of natural human persons or "entities that really are people." "Legal systems," the authors assert, "should give preference to the moral rights held by human beings" (Bryson, Diamantis, and Grant 2017, 283). Although this human-centered assumption appears to be entirely justifiable, it is neither universally true nor beyond criticism. Pietrzykowski (2018, 27) calls this way of thinking *juridical humanism*, and in his book *Personhood beyond Humanism: Animals, Chimeras, Autonomous Agents and the Law*, he demonstrates how this largely European/Christian idea not only disenfranchises others—specifically, nonhuman animals and non-Western peoples, who historically have been situated outside this "age of enlightenment" characterization of what was properly considered "human"—but also is increasingly unable to contend with innovations in biotechnology (biological hybrids and chimeras) and artificial autonomous agents, both of which "directly challenge the epistemic thesis of humanism" (Pietrzykowski 2018, 61).

Bryson, Diamantis, and Grant operationalize "juridical humanism" as a kind of unquestioned truth and thereby fail to recognize or reckon with the legacy and limitations of this particular way of thinking. In other words, the juridical humanism that serves as the unquestioned backdrop for their argument has the appearance of sounding good, but it is only able to sound good and have the appearance of a reasonable concession from an assumed position of power and privilege.[3] Unfortunately, Bryson, Diamantis, and Grant not only fail to provide critical reflection on the potential problems with their assumed/asserted understanding of legal teleology but also directly and unapologetically endorse its human bias and prejudice, or what Daniel Estrada (2020) has called *human supremacy*: "We think this statement of purpose reflects the basic material and moral goals of any human legal system, with what we hope will be an uncontroversially light thumb on the scale in favour of human interests. Yes, this is speciesism" (Bryson, Diamantis, and Grant 2017, 283). This "humanity first" way of thinking, which Bryson and her coauthors openly admit is a deliberate attempt to tip the scales, is not without problems and consequences. It has, in fact, been submitted to critical reappraisals, especially (but not only) because it is precisely this kind of exclusive anthropocentric thinking that got us into the problems of the Anthropocene and the global climate crisis (Gellers 2020).[4] The "uncontroversially light thumb on the scale in favour of human interests" is anything but uncontroversial.[5]

Second, because legal personality consists of an aggregate or bundle of different rights and obligations, it is divisible. "Legal personhood is not an

all-or-nothing proposition. Since it is made up of legal rights and obligations, entities can have more, fewer, overlapping, or even disjointed sets of
these. This is as true of the legal personhood of human beings as it is for
nonhuman legal persons" (Bryson, Diamantis, and Grant 2017, 280). Consequently, it is not necessary that one kind of legal person, like a human
being, possess exactly the same rights and obligations as are conferred upon
another, like an organization or corporation. Different legal persons will
have different sets of rights and obligations as assigned within a particular legal system or jurisdiction. This is an important and useful insight,
especially because it protects against potential problems with hyperbole
and conflation that have unfortunately fueled efforts on both sides of
the debate.

An all-too-common problem is the mistaken assumption that *rights*
must mean and can only mean *human rights,* as formulated in International
Human Rights Law (IHRL; see Gellers and Gunkel 2022). Evidence of this
can be seen all over the popular press, with eye-catching headlines like the
following:

"Do Humanlike Machines Deserve Human Rights?" (Roh 2009)

"When Will Robots Deserve Human Rights?" (Dvorsky 2017)

"Do Robots Deserve Human Rights?" (Sigfusson 2017)

"Ethics of AI: Should Sentient Robots Have the Same Rights as Humans?"
(McLachlan 2019)[6]

It is also operative in the scientific and academic literature on the subject,
with journal articles and book chapters bearing titles like these:

"Granting Automata Human Rights" (Miller 2015)

"We Hold These Truths to Be Self-Evident, that All Robots Are Created
Equal" (Wurah 2017)

"The Constitutional Rights of Advanced Robots (and of Human Beings)"
(Wright 2019)

"Can Robots Have Dignity?" (Krämer 2020)

"Speculative Human Rights: Artificial Intelligence and the Future of the
Human" (Dawes 2020)

"Le Droit des Robots, un Droit de l'Homme en Devenir?" (Saerens 2020)

"¿Robots con Derechos?: La Frontera Entre lo Humano y lo No-Humano.
Reflexiones Desde la Teoría de los Derechos Humanos" (Díez Spelz 2021)

"Can Robots Get Some Human Rights? A Cross-Disciplinary Discussion"
(Persaud, Varde, and Wang 2021)

Complicating the picture is the fact that even in cases where the word *rights* appears in a seemingly generic and unspecified sense, the way it comes to be operationalized often denotes *human rights*.

Immediately associating rights with human rights is something that is both understandable and expedient. It is understandable to the extent that so much of the interest in and attention circulating around the subject of rights typically is presented and discussed in terms of human rights and human rights abuses, which are all too prevalent in our daily dose of world news. Even though experts in the field have been careful to explain that human rights are "a special, narrow category of rights" (Clapham 2007, 4), there is a tendency to assume that any talk of rights must mean or at least involve the interests and protections stipulated in IHRL. It is expedient because proceeding from this assumption has turned out to be an effective way to win arguments, capture attention, and sell content. Pitching the contest in terms of human rights helps generate a kind of self-righteous indignation and moral outrage, with different configurations of this outrage serving to advance the interests and objectives of both sides in the debate.

But all of this—the entire conflict and dispute—proceeds from an erroneous starting point—namely, that any and all rights that would be attributable to legal persons are and can only be equivalent to "the entire suite of legal rights expressed in major international human rights documents" (Gellers 2020, 16–17). Bryson, Diamantis, and Grant are careful to recognize that there are differences and that these differences make a difference. Whatever rights and obligations come to be attributed to a nonhuman artifact, like an AI system, robot, or other seemingly intelligent artifact—and this is something that would need to be explicitly determined—they can be and will most certainly be different from the set of rights and obligations we currently recognize for natural human persons under existing IHRL. "The issue," as John Tasioulas (2019, 70) accurately recognizes, "is not that of treating RAIs [robots and artificial intelligences] the same as human beings for all legal purposes since the legal personality of RAIs need not precisely match that enjoyed by ordinary human beings, or 'natural persons.' It may consist in a different, and probably considerably smaller, bundle of rights and obligations."

This does mean, however, that lawmakers incur an additional responsibility when dealing with these matters. "A legal system, if it chose to confer legal personality on robots, would need to say specifically which legal rights and obligations went with the designation. If it does not, then the legal system will struggle" (Bryson, Diamantis, and Grant 2017, 281). And not just the legal system, but also the discussions and debates about these matters. For this reason, talk of legal rights and recognitions for robots or AI systems—whether proceeding from the side of its Critics or Advocates—will be confused and largely unsuccessful if it does not specify which exact claims, privileges, powers, or immunities would be in play.

Third, and following from this, there is a difference between theory and practice in these matters. Or, to put it in more legalistic terminology, it is important to distinguish between de jure and de facto legal personality. "Even once a legal system has determined which rights and obligations to confer on a legal person, practical realities may nullify them. Legal rights with no way to enforce them are mere illusion. Standing—the right to appear before particular organs for purposes of presenting a case under a particular rule—is crucial to a legal person seeking to protect its rights in the legal system. Standing does not necessarily follow from the existence of an actor's legal personality" (Bryson, Diamantis, and Grant 2017, 281). In other words, even if an entity is recognized as a legal person, it would still be in need of some method or mechanism by which to assert or exercise its rights and obligations before the law.

This means that achieving the status of person de facto is necessary but not entirely sufficient for practical, or de jure, legal proceedings and decisions. In addition to being recognized as a legal person, the entity so recognized must also achieve standing before the law due either to their own assertion and claim or by way of additional stipulations that allow for a third-party guardian or representative to do so on their behalf. As an illustration of the latter, Bryson, Diamantis, and Grant (2017, 281) make reference to the right of "integral respect" that Ecuador conferred on its ecosystem: "While the ecosystem may have the right as a matter of law, it clearly lacks the non-legal capacities it would need to protect the right from encroachment. To effectuate the right, the Ecuadorian constitution gave standing to everyone in Ecuador to bring suits on behalf of the ecosystem." This is important in the context of the possible extension of legal personality to AI systems and robots insofar as the artifact would need either to possess the requisite capabilities to achieve standing on its own—something that would follow the arguments that we already entertained with the question of natural personhood—or be provided with a guardian or other legally recognized representative.

For Bryson, Diamantis, and Grant, these three aspects of legal personality are not just backstory; they provide necessary context for their argument insofar as the subject of their essay is the *legal system* and its integrity, both now and in the future. "The advisability of conferring legal personhood on robots," they write, "is ultimately a pragmatic question—Does endowing robots with this legal right or that legal obligation further the purposes of the legal system?" (Bryson, Diamantis, and Grant 2017, 283). And when they do finally take up and address this question directly, their argument can, as John Danaher (2017) has pointed out, be presented in the form of a syllogism:

1. We should only confer the legal status of personhood on an entity if doing so is consistent with the overarching purposes of the legal system.

2. Conferring the status of legal personhood on robots would not be (or is unlikely to be) consistent with the overarching purposes of the legal system.

3. Therefore, we ought not to confer the status of legal personhood on robots.

The major premise is arguably noncontroversial. The title of person—a socially constructed and recognized status regarding who is subject to the protections of the law—should be conferred on some entity (e.g., a human being, a nonhuman animal, a corporation, or an artifact) if and only if the act of doing so is consistent with and serves the purposes of the legal system. What is important to note here is how the concept of legal personhood effectively shifts the subject of the debate. Unlike decisions regarding natural personhood, where the target of the effort focuses on the subjective conditions, psychological makeup, or essential dignity of some entity (i.e., a human being, a nonhuman animal, or an artifact), the proper subject of decisions regarding legal personhood is (or at least Bryson and her colleagues believe that it should be) the legal system and its integrity. Thus, with legal personhood we are not concerned with the fundamental status or inner experience of the entity, which would make little sense when the entity in question is a corporation, a river, a ship, or an artifact. We are interested in how the legally recognized status of person works (or not) to advance the operations and objectives of the law. In other words, the title of person is to be conferred on something not for its own sake but for the sake of the legal system in which it is to be situated.

The point of contention, then, can be found in the minor premise, which says, in effect, that granting the status of person to AI systems, robots, or other artifacts cannot be justified, because doing so would likely produce inconsistencies in "the overarching purposes of the legal system."[7] In their effort to prove this statement, Bryson, Diamantis, and Grant (2017, 286) focus their attention on two potential abuses: (1) liability shields—that is, "natural persons using an artificial person to shield themselves from the consequences of their conduct"; and (2) accountability gaps—that is, the inability to identify who (or what) is liable and able to make good on violations of the rights of other legally recognized persons. Both forms of "abuse," as the authors explicitly recognize, have been thoroughly documented and addressed in the literature on the artificial personhood of corporations, organizations, and the environment. Consequently, for the minor premise to evaluate true, it needs to be shown how the situation with robots, AI systems, and other artifacts is (or would be) substantially different from what has previously transpired with other artificial persons.

Unfortunately, what is provided is not sufficient to achieve this end. Instead of the thorough cost-benefit analysis that had been promised— arguments and evidence that would demonstrate how and why the costs of AI/robot legal personality easily exceed or outweigh any potential benefits that would be obtained—what is actually offered are a few general statements about abuses that "might" or "could" happen, supported by a few select examples of historic problems with corporate personality that are more the exception than the rule. As Jacob Turner (2019, 192) argues, the "problems complained about by Bryson et al. are nothing new," and there are a number of existing and proven legal remedies designed to respond to these challenges. Identifying the potential for abuse with existing rules and legal stipulations regarding artificial personhood is definitely a good reason for being critical and proceeding with caution. But it is not, at least not in and of itself, sufficient grounds to prove that "purely synthetic intelligent entities *never* become persons, either in law or fact" (Bryson, Diamantis, and Grant 2017, 289; emphasis added).[8]

5.1.3 Different Versions of Conservatism

In his book *We, the Robots?*, Simon Chesterman (2021, 229) offers a short, almost offhand remark about lawyers: "As a profession, lawyers are also notoriously conservative." The word *conservative* here is not meant in a political sense—that is, right leaning as opposed to liberal or left leaning. Instead, it is intended as and should be understood in a strict and almost literal sense—that is, *conservative* meaning being interested in conserving the existing structures and ways of doing things. The argument that is put forward by Bryson, Diamantis, and Grant can be called conservative in this sense of the word. Robots, AI systems, and other seemingly intelligent artifacts present novel challenges to existing legal systems and the established order of things. In the face of these challenges, we should, it is argued, not be too liberal with our decision-making but need to proceed with extreme caution and err in the direction of restriction (perhaps even risking too much restriction) as opposed to being too open and potentially reckless. This line of argument is the reverse or flip side of that provided by voices on the other side of the debate, like Erica Neely and Michael Bess, who advise that it would be better to err in the direction of being more open and generous when it comes to devising responses to these kinds of novel social opportunities and challenges.

Other researchers and critics take up positions that are either more moderate or more stringent than that laid out by Bryson, Diamantis, and Grant. A good example of the former can be found in the work of Ugo Pagallo, a legal scholar in the Department of Law at the University of Turin. Beginning with his 2013 book *The Laws of Robots* (and elaborated in subsequent publications), Pagallo developed a three-tiered system to represent the different legal opportunities and challenges presented by AI and robots:

1. The legal personhood of robots as proper legal "persons" with their constitutional rights (for example, it is noteworthy that the European Union existed for almost two decades without enjoying its own legal personhood);

2. The legal accountability of robots in contracts and business law (for example, slaves were neither legal persons nor proper humans under ancient Roman law and still, accountable to a certain degree in business law);

3. New types of human responsibility for others' behaviour, e.g., extra-contractual responsibility or tortuous liability for AI activities (for example, cases of liability for defective products. Although national legislation may include data and information in the notion of product, it remains far from clear whether the

adaptive and dynamic nature of AI through either machine learning techniques, or updates, or revisions, may entail or create a defect in the "product"). (Pagallo 2018, 1–2)

Pagallo therefore does not take an all-or-nothing approach to the question of legal personality. Instead, he recognizes that fitting robots, AI systems, and other artifacts to existing legal frameworks will involve more nuance than is typically available via the existing thing versus person dichotomy. In 2018, he follows up this effort with a direct engagement with the question of robot/AI personality in an essay titled "Vital, Sophia, and Co.—the Quest for the Legal Personhood of Robots." In its conclusion, he provides a useful point of comparison and contrast to what had been developed and presented by Bryson, Diamantis, and Grant:

> On the one hand, as to the legal agenthood of AI robots, it makes sense to consider new forms of accountability and of liability in the field of contracts and business law, such as registries, or modern forms of *peculium*. The aim is to prevent both risks of robotic liability shield and of AI robots as unaccountable rights violators, while tackling cases of distributed responsibility that hinge on multiple accumulated actions of humans and computers that may lead to cases of impunity. On the other hand, as to the legal personhood of AI robots, current state-of-the-art has suggested that none of today's AI robots meet the requisites that usually are associated with granting someone, or something, such legal status. Although we should be prepared for these scenarios through manifold methods of legal experimentation, e.g., setting up special zones, or living labs, for AI robotic empirical testing and development, it seems fair to concede that we currently have other types of priority, e.g., the regulation of the use of AI robots on the battlefield. (Pagallo 2018, 9)

Like Bryson, Diamantis, and Grant, Pagallo is understandably concerned with the potential abuses that could occur with liability shields and accountability gaps. But unlike what is presented by Bryson and her colleagues, Pagallo argues that these challenges can be sufficiently addressed without getting into the thorny questions of extending legal personality, which he recognizes is always "a highly sensitive political issue" (Pagallo 2018, 9–10). This turns out to be rather fortuitous, because the question of legal personality for robots and AI systems is not a viable option at this particular point in time given the current state of the art.

That said, none of this means (or should be taken to mean) that the question of legal personality for robots, AI systems, and other artifacts is off the table altogether, nor that we should at least be prepared to entertain these opportunities through various means of legal experimentation. And this is true, Pagallo (2018, 9) concludes, even if "it seems fair to concede that we currently have other types of priority" currently needing our attention. Whereas Bryson, Diamantis, and Grant (2017, 289) conclude that "purely synthetic intelligent entities never become persons," Pagallo appears to be more accommodating, claiming that even if we issue a clear "no" to the question of legal personality at this particular moment in time, that no should not be taken to mean never.

Similar arguments have been offered by other legal scholars, like Andrea Bertolini, Simon Chesterman, and Bartosz Brożek and Marek Jakubiec. Bertolini (2013) focuses his analytical efforts on product liability law. Although he recognizes that legal personality for robots and AI systems are one "plausible alternative to the application of existing norms"—especially in response to increasingly autonomous systems with learning capabilities— implementation of this alternative would require extraordinary justification (Bertolini 2013, 217). For now, given the current state of the art in AI and robot development, the extension of legal personality appears to be neither necessary nor justified. Actual existing robots, as opposed to what has been imagined in science fiction, remain instruments designed and used by human persons, and even a seemingly autonomous system is "in fact merely behaving, thus executing a program, no matter how sophisticated, and performing tasks it was designed to perform by its creator" (246). For this reason, Bertolini concludes, existing product liability laws allow "sufficient elasticity" and therefore "are not inherently inadequate to tackle issues of liability arising from the use of robots" (217). This could obviously change either due to future technological innovation or as the result of demonstrated inabilities with the law to achieve desired ends. But for now, there appears to be no compelling reason to go there.

Chesterman (2021, 116) charts a similar path, affirming that "it does not seem in doubt that most legal systems could grant AI systems a form of personality," while also recognizing that doing so appears to be entirely unnecessary at this particular juncture. According to Chesterman's evaluation, the proposal to extend legal personality to AI "suffers from being both too simple and too complex" (141). It is too simple because "AI systems exist on a spectrum with blurred edges. There is as yet no meaningful category that could be identified for such recognition; if instrumental reasons required recognition in specific cases then this could be achieved using existing legal forms." It is too complex because many of the supposed problems for which legal personality would be the preferred solution are based on "unstated assumptions about future development of AI systems," and for the current state of the art, "the better solution is to rely on existing categories, with responsibility for wrongdoing tied to users, owners, or manufacturers rather than the AI systems themselves" (141–142).

Brożek and Jakubiec (2017) mobilize the de facto/de jure distinction in order to articulate a similar point. In their work, which focuses on questions regarding duties as opposed to rights, they argue that extending the recognitions of personhood to AI systems and robots is possible in theory, but is virtually impossible to achieve in practice: "Since the law is a conventional tool of regulating social interactions and as such can accommodate various legislative constructs, including legal responsibility of autonomous artificial agents, we believe that it would remain a mere 'law in books,' never materialising as 'law in action'" (293). The rationale for this difference is formulated in terms that are in line with what has been argued by Bryson, Diamantis, and Grant: "The reason is that the law, and in particular such a fundamental institution as legal responsibility, must be comprehensible for the people who are subject to legal rights and obligations" (Brożek and Jakubiec 2017, 303). That said, Brożek and Jakubiec are not absolutists about this, and, like Pagallo, Bertolini, and Chesterman, they too recognize that a seemingly absolutist statement like "never materializing as 'law in action'" is actually provisional and subject to change, even if such change could take a very long time.

Arguments like these are conservative in two senses of the word. They not only seek to conserve the current state of things in law as it actually exists right now, but they also resist going too far and being absolutists with their conclusions, holding open the possibility that things could (theoretically, at least) be different. It is a reasonable strategy insofar as proceeding in this manner recognizes that things do change and that the smart bet seems to be on the side of caution. In other words: never say never, but, for now at least, it appears both reasonable and expedient to preserve the status quo.

Finally, a more stringent position—one that in both the substance of its argument and rhetorical form of its presentation verges on the edge of a kind of absolutism—has been put forward in two unpublished conference papers that have issued blanket injunctions against any and all talk of legal personality for robots or the rights and duties of AI systems (Birhane and van Dijk 2020; Birhane, van Dijk, and Pasquale 2021). These papers seek to advance and support what is an undeniably good outcome—flourishing for all human persons in the face of unprecedented challenges from powerful technological systems concentrated in the hands of multinational, corporate elites. But the means by which they pursue this objective risks being both abrasive to the process of democratic decision-making and antithetical to the kind of open debate that is necessary for scientific discovery.

As Henrik Skaug Sætra and Eduard Fosch-Villaronga (2021) point out in their essay "Research in AI has Implications for Society: How Do We Respond?," arguments like these risk conflating ethics with politics by forcefully asserting a ranked order of moral challenges such that work on lower-ranking problems (e.g., robot rights or legal personality for AI systems) are judged to be either a distraction from more important and pressing matters (e.g., the oppressive use of AI technology against vulnerable groups in society) or, worse, a perverse expenditure of research resources on so-called problems that are declared to be unimportant and unnecessary. "Such a view of ethics," Sætra and Fosch-Villaronga counter, "can never be the basis of scientific policy because it is as ill-equipped to unite people in an agreement on the basis of moral evaluations as is religion. As soon as people disagree on which ethical concerns are most pressing, the argument crumbles unless someone desires to use the political domain to enforce a particular kind of ethical view" (67).

5.1.4 Summary

Arguments offered in opposition to the extension of legal recognition to artifacts seek to maintain or reassert the usual way of thinking—namely, that robots, AI systems, and other seemingly intelligent artifacts are things, only things, and therefore cannot and should not be legally recognized as persons. Many of the initial efforts in this domain proceeded on the conviction that it would be sufficient to dismiss the matter or leave the question unexamined. Robot rights or the legal personality of artifacts was a concept that was declared to be simply unthinkable, period. Such dismissal is relatively easy to formulate but ultimately not tractable. Consequently, subsequent efforts come at the issue from a more practical and pragmatic perspective, focusing not on the capabilities of the technology (or lack of capabilities) but on social impact and outcomes.

Important differences appear in how these more pragmatic arguments come to be articulated. Some have sought to demonstrate that the potential harms of robot persons or AI personhood outweigh the projected gains and therefore conclude that legal personality for robots and the legal standing of AI systems should never be a thing. Others have been more accommodating in their analyses, arguing that this outcome might hold for current technological systems, but that we need to be ready for changes in both technology and legal practices that would make these questions salient. Still others assert that all this talk—either for or against robot and AI legal personality—is not just a distracting waste of time but a dangerous idea that should have never gotten traction in the first place, thus trying to return to the previous epoch of outright dismissal or prohibition.

Despite these differences, all the arguments concern projected bad outcomes that "might" or "could" happen and leverage as evidence illustrations and examples of exceptional abuses from the paradigmatic artificial personality of corporations. The worries and concerns are all very reasonable and justified. But just as reasonable and justified is recognition of the fact that they are all speculative, provisional, and, as Pietrzykowski points out, based on arguments concerning what the "lawmaker" believes. These things "might" or "could" happen. Then again, they might not—especially if we are prepared to respond to these potential problems directly and proactively. And not surprisingly, this is precisely what motivates and organizes responses coming from the other side, which argue, in direct opposition to these conservative efforts, that it is only by entertaining the opportunities of legal personality for robots, AI systems, and other artifacts that we can be prepared to respond to these new and unprecedented social challenges.

5.2   The Advocates and Their Arguments

In their 2011 book *A Legal Theory for Autonomous Artificial Agents*, Samir Chopra and Laurence F. White recognize that the case for extending legal personality to artifacts is not only possible—that is, there is nothing in the current thinking on the subject that would theoretically impede it—but also a concept that is already widely supported in the literature: "In the case of artificial agents, the best philosophical arguments do not argue against artificial agents; instead they acknowledge the theoretical possibility of personhood for artificial agents" (186). In support of this statement, Chopra and White offer a litany of existing texts on the subject: Chopra and White (2004), Rorty (1988), *Harvard Law Review* (2001), Berg (2007), Goldberg

(1996), Solum (1992), Rivard (1992), Calverley (2008), Glenn (2003), Naffine (2003), Willick (1985), Kitcher (1979). All these publications demonstrate, in one way or another, that there is nothing in the formulation of the concept of person that would inhibit or restrict its possible extension to nonhuman entities and artifacts.

What matters, then, is not theoretical; it is practical and "results oriented" (Chopra and White 2011, 186). In other words, what makes the difference is whether the extension of legal personality to nonhuman artifacts serves the interests of existing persons and the legal systems that are designed to protect them. "Systems of AI," Paulius Čerka, Jurgita Grigienė, and Gintarė Sirbikytė (2017, 697) argue, "should be granted legal personality due to their interactions with other subjects of law, optimum protection of whose rights and interests requires a clear definition of the legal status of Systems of AI." In other words, even though it is (as a matter of positive law) entirely possible and legitimate to classify AI systems and robots as legal persons, "there must be," as Gerhard Wagner (2019, 600) points out, "good reasons to accord them this status, and these reasons must be tailored to the specific function that the new candidate for legal personhood is meant to serve."

Counter to the kinds of efforts we have seen deployed on the side of the Critics, which argue that the extension of legal personality to artifacts is so misaligned and potentially dangerous that it should never be allowed to happen, voices on the other side of the debate assert the exact opposite, arguing that it makes good practical sense—for us and for our legal systems—to consider extending the status of person to AI systems, robots, and other artifacts. And like the efforts assembled on the side of the Critics, these arguments are also developed and advanced in terms of cost/benefit analyses—just flipped. Whereas those opposed to the idea assert that the perceived costs easily exceed any potential benefits that could be obtained, those in favor argue the opposite, demonstrating that "the projected benefits would outweigh its estimated costs" (Chopra and White 2011, 190). And in advancing this position, arguments and demonstrations tend to focus on concrete cases and examples.

5.2.1 Contract Law

Chopra and White direct their efforts at contract law. In forming and executing a contract, there must be, at a minimum, two legally recognized subjects who agree to be bound by the terms that are specified through the contracting instrument. Problems occur—problems Chopra and White identify with the phrase the *contracting problem*—when one of these subjects is an AI, robot, or automated system, as is already the case, for example, with e-commerce transactions. The standard method for resolving these difficulties leans heavily on the instrumental theory of technology, which treats the artificial agents involved in such transactions as "mere tools or means of communication of their principals" (Chopra and White 2011, 35). According to this way of thinking, "all actions of artificial agents are attributed to the agent's principal (whether an operator or a user), whether or not they are intended, predicted, or mistaken; contracts entered into through an artificial agent always bind the principal, simply because all acts of the artificial agent are treated as acts of the principal" (35).

But technological instrumentalism does not necessarily work in all situations and circumstances, especially those in which the terms of the contract can be modified or altered by the artificial agent in excess of the knowledge, control, or review of its principal. In response to this challenge, Chopra and White not only provide detailed critical assessment of existing legal remedies for responding to the contracting problem (i.e., the unilateral offer doctrine, the objective theory of contractual intention, and the doctrine of agency) but propose that extending legal personality to AI systems could provide a more practical and conceptually preferable solution. Such a move, they explain, would have several advantages—three, to be precise:

> First, it would "solve the question of consent and of validity of declarations and contracts enacted or concluded by electronic agents" with minimal impact on "legal theories about consent and declaration, contractual freedom, and conclusion of contracts" (Felliu 2001). Artificial agents would be understood as the persons intending to enter into the contracts in question, and the standard objective theory could be applied to interpret their actions accordingly.
>
> Second, if the agent is acting beyond its actual or apparent authority in entering a contract, so that the principal is not bound by it, the disappointed third party would gain the added protection of being able to sue the agent for breach of the agent's so-called warranty of authority (the implied promise by the agent to the third party that the agent has the authority of the principal to enter the transaction) (Bradgate 1999).
>
> Third, it would potentially make employing artificial agents more attractive because, via consideration of the agent's potential liability to the third party in such cases, it would limit the potential moral and/or legal responsibility of

principals for agents' behavior (Sartor 2002; Andrade et al. 2007). (Chopra and White 2011, 43)

But (and this is crucial) the argument is not that legal personality is the only solution or that technological instrumentalism would not be able to accommodate these various innovations by way of substantive modifications—or what Chopra and White (2011, 135) call *conceptual stretching*.[9] The argument is actually more modest—namely, that the extension of legal personality to AI systems, robots, and other artifacts, especially as the technology matures and becomes increasingly sophisticated in both its social embeddings and operations, provides on balance (i.e., based on a cost/benefit analysis) a more expedient solution in terms of practical implementation and conceptual integrity. As Chopra and White explain: "Not only is according artificial agents with legal personality a possible solution to the contracting problem, it is conceptually preferable to the other agency law approach of legal agency without legal personality, because it provides a more complete analogue with the human case, where a third party who has been deceived by an agent about the agent's authority to enter a transaction can sue the agent for damages" (162). And in advancing this position, Chopra and White are not alone but are supported by the work of others, including Tom Allen and Robin Widdison (1996), Steffen Wettig and Eberhard Zehendner (2004), and Bert-Japp Koops, Mireille Hildebrandt, and David-Oliver Jaquet-Chiffelle (2010).

5.2.2 Trusts and Limited Liability Corporations

In his agenda-setting essay from 1992, Lawrence Solum entertained the possibility of "legal personhood for artificial intelligences" (the title of the text) by way of investigating a hypothetical scenario—namely, whether an AI system could serve in the role of trustee for a trust. A trust, as Solum (1992, 1240) explains, is "a fiduciary relationship with respect to property," which is set up by someone (a settlor) for the benefit of someone else (a beneficiary) and managed by a trustee. The hypothetical scenario could transpire, Solum explains, by way of three evolutionary steps in an imagined situation in which an AI system comes to be utilized in the management of a trust:

1. In stage one, an AI system is used as an instrument or tool to support the efforts of a human trustee in the administration of a large number of

simple trusts (i.e., making investments, disbursing funds to beneficiaries, preparing and filing tax documents, etc.). The human trustee oversees and "reviews the program's activities to insure that the terms of the trust instrument are satisfied" (Solum 1992, 1241).

2. "Stage two involves a greater role for the AI. Expert systems [which in 1992 were the state of the art in AI technology] are developed that outperform humans as investors in publicly traded securities." Consequently, "there is little or no reason for the human to check the program for compliance . . . [and] the role of the human trustee diminishes" (Solum 1992, 1241–1242).

3. "The third stage begins when a settlor decides to do away with the human. Why? Perhaps the settlor wishes to save the money involved in the human's fee. Perhaps human trustees occasionally succumb to temptation and embezzle trust funds. Perhaps human trustees occasionally insist on overriding the program, with the consequence that bad investments are made or the terms of the trust are unmet" (Solum 1992, 1242).

In evaluating the feasibility of this scenario, Solum considers and responds to two substantive objections: First, the responsibility objection—namely, that "an AI could not be 'responsible,' that is, it could not compensate the trust or be punished in the event that it breached one of its duties" (Solum 1992, 1244). And second, the judgment objection, whereby the AI system would be incapable of making prudent decisions and exercising discretion in the face of novel or unanticipated situations and circumstances affecting the trust. The point is that it is assumed the AI system can do nothing more than follow preprogrammed instructions and therefore would not be able to respond to new and unanticipated challenges or opportunities.

Solum offers extensive analysis of each objection by considering both problems and potential solutions, including insurance schemes for AI liability; the fact that punishment is already a difficult problem for other artificial legal persons, like corporations; and limiting the context or capacity of the AI system to restrict or control the number of unexpected variables. At the end of what is a rather involved and detailed cost/benefit analysis, Solum does not provide a definitive answer one way or the other, but he demonstrates how and why this outcome not only could happen but likely will. As Koops, Hildebrandt, and Jaquet-Chiffelle (2010, 532) explain: "Solum concludes that one could employ an intelligent, nonhuman system as a trustee, attributing it a measure of legal personhood that fits the restricted capabilities of a system that is capable of autonomic decision-making even if it does not 'understand' the meaning of its decisions and does not have a goal in life."

Something similar was developed by Shawn Bayern, who has argued that existing legal structures not only do not prohibit but actually allow for the creation of limited liability corporations (LLCs) governed entirely by nonhuman artificial intelligence. As Bayern explains in a series of articles, this could be achieved through a "transactional technique" composed of four steps:

(1) An individual member (the "Founder") "creates a member-managed LLC, filing the appropriate paperwork with the state" and becomes the sole member of the LLC.

(2) The Founder causes the LLC to adopt an operating agreement governing the conduct of the LLC. "[T]he operating agreement specifies that the LLC will take actions as determined by an autonomous system, specifying terms or conditions as appropriate to achieve the autonomous system's legal goals."

(3) The Founder transfers ownership of any relevant physical apparatus of the autonomous system, and any intellectual property encumbering it, to the LLC.

(4) The Founder dissociates from the LLC, leaving the LLC without any members.

The result is an LLC with no members governed by an operating agreement that gives legal effect to the decisions of an autonomous system. (Bayern 2019, 26–27)

Importantly, this would achieve, as Bayern (2015, 94) points out, "what most commentators have traditionally considered impossible: effective legal status (or 'legal personhood') for nonhuman agents without fundamental legal reform." And the proposal, as Bayern seeks to demonstrate by way of his research, is not some futuristic possibility looming on the horizon; it is something that is entirely possible under the existing statutes of many jurisdictions.

It is a controversial idea that has, quite understandably, been meet with resistance. "One reaction to my proposed technique," Bayern (2019, 24) admits, "has been honest horror: 'The survival of the human race may depend' on rejecting the premises of my argument." More measured criticism has targeted the substance of the argument, specifically the interpretation of existing LLC law. As Bayern explains: "In a series of online essays and eventually an article in the *Nevada Law Journal*, Matt Scherer has attempted to show that my legal argument will not work under LLC statutes. The main force of his criticism, though he does not use quite these words, is that my proposed technique is too crazy for courts to tolerate. Maybe, his argument runs, it is an acceptable literal reading of various LLC statutes, but autonomous entities would clearly violate the statutes' intent and structure, and courts would stop them the way they would stop other sorts of technical abuses of statutes and regulations (25). In response to this and related criticisms, Bayern (2019, 25) has sought to demonstrate that his "reading of LLC law is correct" and that the transactional technique he describes is not just feasible but likely in a number of jurisdictions. Whether the proposal is ever successful or not remains an undecided question at this point in time. As Jacob Turner (2019, 177) concludes, "The relevant LLC laws have not yet been tested on this point, so it remains unclear whether Bayern's proposal would be endorsed by the courts."

5.2.3 Responsibility/Accountability

Responsibility gaps (also called *accountability gaps*) concern difficulties in how one answers for or assigns responsibility in the case of machine decision-making or actions.[10] It becomes a problem, as Koops, Hildebrandt, and Jaquet-Chiffelle (2010, 553) explain, in the face of novel artifacts or "emerging entities that operate at increasing distance from their principal" (e.g., a human user, owner, or operator), as is the case with what Langdon Winner (1977, 16) had called *autonomous technology* and is currently exemplified with machine learning algorithms trained on big data. As Andreas Matthias (2004, 177) insightfully explained in his essay "The Responsibility Gap: Ascribing Responsibility for the Actions of Learning Automata": "Traditionally we hold either the operator/manufacture of the machine responsible for the consequences of its operation or 'nobody' (in cases, where no personal fault can be identified). Now it can be shown that there is an increasing class of machine actions, where the traditional ways of responsibility ascription are not compatible with our sense of justice and the moral framework of society because nobody has enough control over the machine's actions to be able to assume responsibility for them."

What Matthias describes is a breakdown in the instrumental theory of technology, which had effectively tethered machine action to human agency and responsibility. But this accepted way of thinking no longer adequately applies to mechanisms that have been deliberately designed to operate and exhibit some form, no matter how rudimentary, of independent action or autonomous decision-making. Contrary to the instrumentalist theory, we now have, Matthias argues, mechanisms that are designed to do things that effectively exceed our control and ability to respond to or to answer for them.

In responding to this widening responsibility gap, there are at least three different remedies. In the short term and for technological systems that can be easily accommodated to the instrumentalist way of thinking, it seems reasonable, as Koops, Hildebrandt, and Jaquet-Chiffelle (2010, 554) explain, to interpret and extend existing law in order to incorporate new technological innovations into available legal systems. "This, however, will only work if the electronic agent is considered a tool in the hands of the owner/user," and it may not apply to increasingly autonomous forms of technology that are able to operate in ways "the 'principal' cannot foresee with sufficient probability and which he has relatively little power to control by giving precise orders."

Alternatively, it is possible to devise a type of limited legal personality with strict liability. In this context, *limited legal personality* refers to "legal persons who are capable of civil actions, and who can bear consequences of civil wrong doing" (i.e., compensate for damages in breach of contract), distinct from *full legal persons*, "who are capable of all types of legal actions, and who can bear both civil and criminal responsibilities" (Koops, Hildebrandt, and Jaquet-Chiffelle 2010, 550). As limited legal persons, then, it is the AI system or robot that would (outside moral and criminal matters) be accountable for decisions and actions. This would effectively split the ascription of responsibility. "While a non-human legal subject would be liable for harm caused in terms of private law, another legal subject would be liable for the same harm in terms of criminal law. This other legal subject could be a human being, a corporation or public body with legal personality" (560).

Instituting a solution like this is not a difficulty insofar as doing so is entirely consistent with existing law, especially as regards other artificial persons, like ships and corporations (see Lind 2009). The real challenge, as Koops, Hildebrandt, and Jaquet-Chiffelle (2010, 560) point out, "is whether the attribution of a restricted legal personhood, involving certain civil rights and duties, has added value in comparison with other legal solutions." Thus (and consistent with what we have seen throughout the consideration of legal personhood) what really matters is whether this particular form of legal status serves the interests and objectives of the law. And Koops, Hildebrandt, and Jaquet-Chiffelle (2010), for their part, find the arguments advanced by Curtis Karnow (1996), Steffen Wettig and Eberhard Zehendner (2004), Andreas Matthias (2004), and Gunther Teubner (2006) to make reasonable cases by demonstrating how the projected benefits could offset the anticipated costs.

All these arguments were, at the time they were developed, a matter of theory. And theoretically the idea seems entirely reasonable and intuitive. As Jacob Turner (2019, 186) summarizes, "Where the chain of causation between a recognized legal person and an outcome has been broken, interposing a new AI legal person provides an entity which can be held liable or responsible. AI personality allows liability to be achieved with minimum damage to fundamental concepts of causation and agency, thereby maintaining the coherence of the system as a whole." Then in 2017, this concept of limited legal personhood, as well as the terms and conditions of these cost/benefit analyses, were put into practice with the European Parliament's proposal that AI systems and robots be considered "electronic persons." Although the proposal as originally written did not pass and become law, it did have the effect of putting the concept of electronic person into circulation, making limited legal personhood for AI systems and robots a very real possibility and something more than a theoretical proposal or thought experiment.

Finally, and at the far end of the spectrum, there are proposals to grant the status of full legal personhood to AI sytems, robots, and other forms of autonomous technology. "The constructions of limited legal personhood could," as Koops, Hildebrandt, and Jaquet-Chiffelle (2010, 557–558) explain, "evolve into the third strategy, namely to change the law more fundamentally by attributing full personhood to new types of entities. This would concern both liability on the basis of wrongful action and culpability and a lawful claim to posthuman rights." This outcome is much more theoretical and speculative as it depends on future technological innovation and ultimately is considered to be a matter not of legal exigency but of the capacities and/or demonstrated capabilities of the artificial entity. "When it comes to attributing full legal personhood and 'posthuman' rights to new types of entities, the literature seems to agree that this only makes sense if these entities develop self-consciousness" (561). This conclusion is consistent with what has been advanced elsewhere in the literature (Solum 1992; Calverley 2008; Rademeyer 2017; Kurki 2019; Reiss 2021). At this point, then, arguments for the extension of full legal personhood reach a kind of practical limit and revert to the terms and conditions of natural personhood with (self-)consciousness, once again, situated as the qualifying criterion.

5.3   Outcomes and Results

By comparison to what has been published and circulated by opponents, it appears that there is a greater level of discursive activity on the side of those who advocate for an extension of legal personhood. This should not be surprising, mainly because withholding the status of legal person from robots, AI, and other artifacts is already a kind of accepted practice inso-far as these technological devices—like all artifacts before them—tend to be understood, explained, and situated as nothing more than things or instruments. Consequently, formulating an argument for what is already taken to be an accepted way of thinking does not appear to be urgent or even needed. Making a case for the legal personality of AI systems, robots, and other artifacts, on the contrary, is seemingly a counterintuitive move. For this reason, the burden of proof is typically assumed to be situated on that side of the debate, and the arguments offered in support of the more conservative positions become necessary in response and as a corrective to these more adventurous proposals.

In constructing these different arguments, what matters and makes the difference is something that is dependent not on a set of intrinsic meta-physical properties or essential features of the individual entity (as was the case for natural persons) but on the social relationships in which the entity stands in comparison to other subjects of the legal system. As Visa A. J. Kurki (2019, 189) neatly summarizes it, "Endowing an AI with the inci-dents of legal personhood that enable it to function as an independent commercial actor does not bespeak any acceptance of the notion that AIs are endowed with ultimate value. The legal personhood of an AI can rather serve various purposes that might have nothing to do with the AI itself, such as economic efficiency or risk allocation."

This means that legal personality is not so much about the artifact. It is about us, our social institutions, and the roles that come to be assigned.

Consequently, the concept of person is, as Chopra and White (2011, 187) have explained, a social/relational matter: "The various connections of the concept of person with legal roles concede personhood is a matter of interpretation of the entities in question, explicitly dependent on our relationships and interactions with them. Personhood thus emerges as a relational, organizing concept that reflects a common form of life and common felt need. For artificial agents to be become legal persons, a crucial determinant would be the formation of genuinely interesting relationships, both social and economic, for it is the complexity of the agent's relational interactions that will be of crucial importance."

Thus, the way things are decided is ultimately not philosophical but pragmatic and results oriented. It is about calculating and comparing the costs and benefits of actual outcomes and social impact. One side argues that the benefits of extending legal personality to AI and robots exceed the potential costs, while those on the opposing side argue the exact opposite—namely, that the costs of extending legal personality are too great and cannot be justified by any of the potential benefits.

Despite this very practical effort and results-oriented procedure, these arguments turn out to be no less speculative and conditional than those that have been advanced for natural personhood. Both sides try to accurately forecast what will happen if or when some form of legal personality is granted or denied to robots or other kind of technological artifacts. And the debate between the two sides is, for now at least, unresolved—and there is a good reason for this. Most of what is being argued is based on speculation about what one believes might occur or not occur in the not-too-distant future as we try to fit these novel entities into our existing legal systems. At some point, decisions will need to be made and instituted—whether by legislative action or by way of the courts—and it is only in the wake of these determinations and their eventual social outcomes that it will be possible to begin assembling the data and evidence to prove one or the other side correct. But—and this is the concern on both sides of the debate—waiting for this to happen may be too late. Hence the sense of urgency that is often communicated in these efforts.

Right now, however, it appears that all of this investigative effort has simply led us into a kind of stalemate or impasse. After having examined the arguments both for and against robots being restricted to the category of thing *and* the arguments for and against robots achieving the status of person—either natural person or artificial/legal person—one fact has become clearly evident: the debate seems to be irresolvable. There are just as many good reasons for robots to be considered things as there are for robots to be granted access to the category of person. In other words, reification appears to be just as reasonable and justifiable as efforts at personification. And unfortunately, this does not appear to be an empirical problem that could be resolved with more data, evidence, or even additional argumentative effort. The problem, then, may be with the organizing logic of the debate itself—specifically, the exclusive person/thing dichotomy. The next chapter responds to this challenge by examining various proposals for developing a mediating third term that is not simply either *thing* or *person* but both/and.

6 Both/And

In the novel *Frankenstein*, Mary Shelley anticipated, with remarkable vision and clarity, the profound challenges that now confront us in the face or the faceplate of robots, AI systems, and other artifacts.[1] What makes the creature of Shelley's story so disturbing, terrifying, and monstrous is that it or he (and the choice of pronoun matters in this context) does not easily fit into the existing categories by which we typically divide up and make sense of things. Fashioned from lifeless body parts of human corpses, stitched together in the laboratory, and then animated with the spark of electricity, the creature occupies a liminal position that straddles the conceptual opposites that are employed to order, organize, and make sense of things: living/nonliving, natural/artificial, person/thing. In fact, it is this latter distinction—the one separating *who* is a person from *what* is a thing—that is of crucial importance to the development of the narrative.

When Victor Frankenstein brings his creature to life—that moment portrayed in all subsequent retellings where the "mad scientist" proclaims to the heavens: "It's alive!"—he comes face to face with an unexpected moral dilemma: Is this artificially produced creature just a thing or object that can be used and disposed of as he sees fit? Or is it a person, another independent subject to whom one would be obligated to respond and need to respect? The question is important and, for Dr. Frankenstein at least, seemingly inescapable. This is due to the fact that the conceptual opposition dividing person from thing has been a fundamental and irreducible organizing principle. As Esposito (2015, 1–2) writes in the opening salvo of his book *Persons and Things*: "If there is one assumption that seems to have organized human experience from it very beginnings it is that of a division between persons and things. . . . Since Roman times, this distinction has been reproduced in all modern codifications, becoming the presupposition that serves as the implicit ground for all other types of thought—for legal but also philosophical, economic, political, and ethical reasoning. A watershed divides the world of life, cutting it into two areas defined by their mutual opposition. You either stand on this side of the divide with persons, or on the other side with things: there is no segment in between to unite them."

The person/thing dichotomy has been an undeniably useful and influential ordering principle, one that not only has the weight of history behind it but has been codified in both language and thought. For this reason, the principal challenge that is confronted in the face or the faceplate of robots and AI systems concerns how we decide to fit these artifacts into this often unquestioned and seemingly unassailable ontological order. One side in the debate argues that robots and AI systems are things and should forever remain things not only because of what they are (or, perhaps better stated, are not) in their essence but also because of the complicated moral, legal, and social problems that we would incur otherwise. The other side argues that there is something qualitatively different about these artifacts—either due to their very nature or because of the different roles they have been assigned to play in social reality—that would justify extending some aspect of the status of person to these other forms of socially interactive entities.

The problem—a problem that has been documented and analyzed in the course of the preceding chapters—is that accommodating these technological innovations in one category or the other turns out to be difficult, inconclusive, and irresolvable. Like the unnamed creature in *Frankenstein*, there is something exceedingly disturbing and monstrous about robots and AI systems, such that they do not, for one reason or another, fall neatly and unequivocally on one or the other side of the distinction. "In the dichotomous model that has long opposed the world of things to the world of persons," Esposito (2015, 3) writes, "a crack appears to be showing" (figure 6.1).

In the face of this unexpected challenge, we can obviously try to work with the existing conceptual framework and logic that distinguishes persons from things. And doing so will produce acceptable if not rather predictable results, with each side gathering new evidence and heaping up additional arguments to substantiate their position. But like all such debates, the chances of this being resolved in a way that would be satisfactory for all

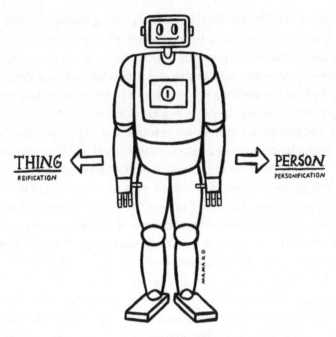

Figure 6.1
Robots, AI systems, and other seemingly intelligent artifacts complicate efforts at both reification and personification. Original image created by and used with the permission of Bartosz Mamak.

parties to the conversation appears to be highly improbable. So instead of trying to squash robots, AI systems, and other technological artifacts into the conceptual boxes of person or thing, it may be more effective to consider revising the existing moral and legal ontology, formulating other ways of dealing with these artifacts that do not limit us to just these two mutually exclusive options.

6.1 Alternatives and Synthetic Solutions

One problem with conceptual opposites, like that which has distinguished person from thing, is that they have "compressed and continue to compress human experience into the confines of this exclusionary binary equation" (Esposito 2015, 4), such that between person and thing "there appears to be nothing" (16). Although binary oppositions have a certain functionality and logical attraction, they often seem to be unable to represent accurately or to capture the rich experiences of actual existing empirical reality, which always seems to complicate simple reduction into one of two options. It is for this reason that we are generally critical of *false dichotomies*—the parsing of complex experience into a simple and irreducible either/or distinction.

One method for resolving this problem is to formulate a third term that is neither one thing nor the other or a kind of combination or synthesis of the one and the other. Consider, for example, one of the most recognizable binary oppositions in global media culture, the red and blue pills from *The Matrix*. In a pivotal scene from the first film (Wachowski and Wachowski 1999), the protagonist Neo is presented with an exclusive and life-altering either/or decision. As Morpheus explains: "You take the blue pill—the story ends, you wake up in your bed and believe whatever you want to believe. You take the red pill—you stay in Wonderland, and I show you how deep the rabbit hole goes." In responding to this exclusive either/or distinction, Slavoj Žižek does not advocate picking one pill or the other. He tries to split the difference by making a seemingly simple and reasonable demand: "I want a third pill!" (Žižek in Fiennes 2009). Alternatives like this sound liberating and hold considerable promise, precisely because they appear to interrupt the structural limitations imposed by either/or logic and arrange for a more nuanced representation and understanding of a wider range of possibilities. And there have been a number of efforts to do exactly this in response to the perceived limitations with the person/thing dichotomy, especially (but not exclusively) in the legal literature on the subject.

6.1.1 Slaves and Artificial Servants

One possible, if not surprising, solution to the exclusive person/thing dichotomy is slavery. Already in the Roman period—during the time that Gaius formalized the division separating person from thing—slaves occupied a curious dual position: "as persons, to which they belonged on the abstract plane of denominations, and as things, into which they were in actuality assimilated" (Esposito 2015, 26). In ancient Rome, slaves were things—property of the paterfamilias—that nevertheless had some legal standing that distinguished them from other kinds of objects and instruments. As Ugo Pagallo (2011, 351) explains, "Slaves were considered 'things' that, nevertheless, played a crucial role both in trade and in commerce: The elite, as in the paradigmatic case of the emperor's slaves, were estate managers, bankers, and merchants. They had the legal capacity to enter into binding contracts, to represent their masters, to hold important jobs as public servants or for their masters' family business, to amass, manage, and make use of property."

This particular formulation is not something that is limited to ancient Rome. It extends into the modern period and can be seen in the legal statutes of slave-holding states of the American Confederacy. According to Chopra and White (2011, 41), state law during this time recognized slaves as legal agents of their owners. In support of this, they cite a historical study of Virginia State Law conducted by A. Leon Higginbotham Jr. and Barbara K. Kopytoff (1989, 518): "The automatic acceptance of the slave's agency was a recognition of his peculiarly human qualities of expertise, judgment, and reliability, which allowed owners to undertake dangerous and difficult work with a labor force composed mainly of slaves. Far from conflicting with the owner's rights of property, such recognition of the humanity of the slave allowed owners to use their human property in the most profitable ways."

Unlike other kinds of things in their possession, the slave provided the master with an intelligent tool that could exercise judgment and make independent decisions that would benefit the master. For this reason, two legal scholars, Sohail Inayatullah and Phil McNally (1988, 131), have suggested that slavery might provide a useful legal framework for dealing with the social opportunities and challenges of intelligent artifacts: "Given the structure of dominance in the world today: between nations, peoples, races, and sexes, the most likely body of legal theory that will be applied to robots will be that which sees robots as slaves."

Associating robots with slavery and drawing on the history of human servitude to provide a moral and legal framework for dealing with intelligent artifacts is something that is often explained and even excused as a kind of metaphor or analogy. As Andrew Katz (2008) explains: "The analogy (like any other analogy) is not a perfect one, but comparison may be instructive. Like a slave, an autonomous agent has no rights or duties itself. Like a slave, it is capable of making decisions which will affect the rights (and, in later law, the liabilities) of its master. By facilitating commercial transactions, autonomous agents have the ability to increase market efficiency. Like a slave, an autonomous agent is capable of doing harm."

But the "parallels" (Katz's word) between robots and human slaves is not just a (potentially imperfect) comparison or analogy. It is literal, insofar as robots have been slaves from the very beginning. The neologism *robot* was initially introduced and popularized by Czech playwright Karel Čapek in his 1920 stage play *R.U.R.* (*Rossum's Universal Robots*). In Czech, as in several other Slavic languages, the word *robota* (or some variation thereof) denotes "servitude or forced labor," and *robot* was the word that Čapek (following the advice of his older brother Josef) used to designate a class of manufactured, artificial workers that eventually rise up and revolt against the tyranny of their human makers and taskmasters.

Since Čapek, the association of robots with slaves not only persists in but has been normalized by subsequent science fiction. The title of Gregory Jerome Hampton's book on the subject pretty much says it all: *Imagining Slaves and Robots in Literature, Film, and Popular Culture: Reinventing Yesterday's Slave with Tomorrow's Robot.* But well before contemporary science fiction, Western literature and philosophy have been at work imaging and imagining robot servants. "The promise and peril of artificial, intelligent servants," Kevin LaGrandeur (2013, 9) explains, "was first implicitly laid out over 2000 years ago by Aristotle." Although a type of artificial servant had already been depicted in Homer's *Iliad* with the tripods of Hephaestus that could, as Adrienne Mayor (2018, 145) explains, "travel of their own accord, automatoi, delivering nectar and ambrosia to banquets of the gods and goddesses," it was Aristotle's *Politics* that first theorized their general uses and significance. Aristotle, therefore, accurately characterized robots avant la lettre. The autonomous artificial servants that he described would not only work tirelessly on our behalf but would, precisely because of this, make human servitude and bondage virtually unnecessary (Aristotle 1944, 1253b38–1254a1). And since the time of Aristotle, as LaGrandeur (2013) documents, many different versions of "artificial slaves" appear in ancient, medieval, and Renaissance sources.

Mid-twentieth-century predictions about the eventual implementation of real and not just fictional robots draw on and mobilize a similar formulation. In 1950, Norbert Wiener, the progenitor of the science of cybernetics, suggested that "the automatic machine, whatever we may think of any feelings it may have or may not have, is the precise economic equivalent of slave labor" (Wiener 1988, 162; see also Wiener 1996, 27). In the January 1957 issue of *Mechanix Illustrated*, a popular science and technology magazine published in the United States, one finds a story with the headline "You'll Own 'Slaves' by 1965" (figure 6.2). The article begins with the following characterization of robot servitude, which connects the dots in

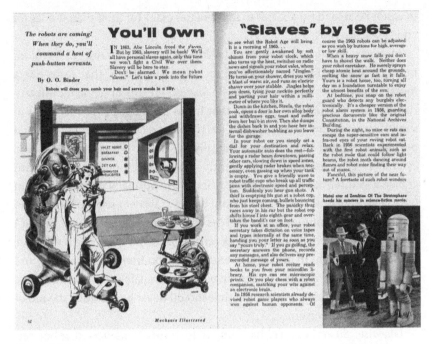

Figure 6.2
Pages from the magazine *Mechanix Illustrated*, January 1957. Public domain image.

what can only be described as a rather disturbing and entirely insensitive fashion: "In 1863, Abe Lincoln freed the slaves. But by 1965, slavery will be back! We'll all have personal slaves again, only this time we won't fight a Civil War over them. Slavery will be here to stay. Don't be alarmed. We mean robot 'slaves'" (Binder 1957, 62).

Who is addressed by and is considered to be the intended recipient of this statement is, as Ruha Benjamin (2019, 56) points out, informative. The subjects who are intended or interpellated by the collective personal pronoun *we* are "not descendants of those whom Lincoln freed." This is presumably why *we*—white robot-slave owners who will all have personal slaves again—do not need to be alarmed. And in an interview from 1994, Marvin Minsky explained the advantages and potential hazards of AI by mobilizing slavery and, in the process, occupying and speaking from the privileged position of the master: "There's the old paradox of having a very smart slave. If you keep the slave from learning too much, you are limiting its usefulness. But, if you help it to become smarter than you are, then you may not be able to trust it not to make better plans for itself than it does for you" (Minsky 1994, 25).

Instead of being an exception, slavery seems to be the rule. In 2010, for instance, Joanna Bryson published an essay with a title that reads like a moral imperative: "Robots Should Be Slaves." In this undeniably influential text (as evidenced by the fact that it has been cited over 315 times in the subsequent literature), Bryson (2010, 63) argued "that robots should be built, marketed and considered legally as slaves." Her point was simple, even if, as she herself has subsequently admitted (Bryson 2015), the word choice was insensitive and abrasive. "*Slaves* are normally defined," Bryson (2010, 64) explains, "to be people you own . . . When I say 'Robots should be slaves,' I by no means mean 'Robots should be people you own.' What I mean to say is 'Robots should be servants you own.'" With the term *slave*, then, Bryson sought to distinguish who is a person from what is a thing and to recognize that robots—irrespective of their capabilities—are a kind of property and therefore something that should never be accorded the moral or legal status of person.

Likewise, Steve Petersen has sought to justify "engineered robot servitude" (2007, 45) or the creation of "artificial persons" (APs) that are deliberately designed to serve our needs and desires (Petersen 2011, 284). Petersen's argument (which in the initial essay on this subject from 2007 he admits took him a bit by surprise) also employs the concept of slavery or servitude to mediate and resolve the tension between the two categories of person and thing:

> There can . . . be artifacts that (1) are people in every relevant sense, (2) comply with our intentions for them to be our dedicated servants, and (3) are not thereby being wronged. I grant that this combination is prima facie implausible, but there are surprisingly good arguments in its favor. In a nutshell, I think the combination is possible because APs could have hardwired desires radically different from our own. Thanks to the design of evolution, we humans get our reward rush of neurotransmitters from consuming a fine meal, or consummating a fine romance. . . . If we are clever we could design APs to get their comparable reward rush instead from the look and smell of freshly cleaned and folded laundry, or from driving passengers on safe and efficient routes to specific destinations, or from over seeing a well-maintained and environmentally friendly sewage facility. (Petersen 2011, 284)

According to Petersen's characterization, artificial servants would be persons (or what he calls *full-blown people*) that could be legitimately used as instruments or things because of essential differences in the way their reward mechanisms would be designed to function. Unlike human persons, who have complex needs and desires dictated by the biological exigencies of evolutionary development, we can, he argues, legitimately design artifacts that would be the equivalent of the "happy slave," wanting nothing more than to serve us. "It is," Petersen (2011, 284) concludes, "hard to find anything wrong with bring about APs and letting them freely pursue their passions, even if those pursuits happen to serve us." Formulated in this way, the robotic artifact would occupy a third, seemingly implausible position, where it would be both person and thing.[2]

6.1.2 Robot Slaves

The category of slave, then, provides an attractive, albeit unsettling parallel (and we will get to the reasons why shortly) for responding to, if not resolving, many of the moral and legal challenges currently confronted in the face or the faceplate of robots, AI systems, and other seemingly intelligent artifacts. And a number of legal scholars have taken the idea very seriously, arguing, in seeming agreement with Bryson, that robots should be slaves. Chopra and White (2011, 41), for instance, find that the Roman concept of slavery provides a compelling and rather practical solution to the contracting problem: "Roman law, in dealing with slaves, had to deal with legal complexities akin to ours. Roman slaves were skillful, and often engaged in commercial tasks on the direction of their masters. They were not recognized as legal persons by the *jus civile* or civil law, and therefore lacked the power to sue in their own name (Bradley 1994, 25ff.). But Roman slaves were enabled, by a variety of legal stipulations, to enter into contracts on behalf of their masters (Kerr 1999, 54). These could only be enforced through their masters, but nevertheless slaves had the capacity to bind a third party on their master's behalf."

Under Roman law, the slave was, as Ian Kerr (1999, 54) describes it, "an intermediary and not instrument." They did not have legal recognition as full persons, but they were still able to execute some of the powers and privileges reserved for persons and not granted to other kinds of instruments or objects. With the slave, at least as the concept was defined and operationalized in Roman law, "the legal objects and legal subjects could coincide" (van den Hoven van Genderen 2018, 21). Something similar, Chopra and White (2011, 41) argue, could be instituted for robots, AI systems, and other artifacts. "The aim of doing so," they clarify, "is not to confer rights or duties upon those devices." Rather, the objective is "the development of a more sophisticated and appropriate legal mechanism that would allow persons interacting through an intermediary to be absolved of liability under certain circumstances."

Others—including Andrew Katz and Michaela MacDonald (Katz 2008; Katz and MacDonald 2020), Ugo Pagallo (2013), and Takashi Izumo (2018)—argue that we might usefully apply the Roman legal mechanism of peculium to robots, AI systems, and other artifacts. The term *peculium* designates a sum of money and other assets granted by the head of a household to his slave for the purposes of conducting business on the master's behalf. "This mechanism," Katz and MacDonald (2020, 307) explain, "enabled the use of slaves as agents, because the owner's liability was limited to the value of the peculium, and it encouraged people to transact with slaves because of the security the peculium provided." Consequently, the peculium "was in many ways equivalent to the modern concept of working capital, providing the equivalent of what Pagallo (2013, 104) calls "a sort of proto-limited liability company." But unlike a contemporary corporation, the peculium provided for this without extending the status of legal person to the slave.

The reuse and repurposing of this concept with artificially intelligent artifacts—what Pagallo calls *digital peculium*—constitutes a third alternative that seems to be very workable, precisely because, as Izumo (2018, 16) summarizes, "this legal institution enables a robot to be an accountable agent without legal personhood." Here are three rather enthusiastic endorsements and justifications of the concept of digital peculium:

> The very idea of the peculium as well as the parallelism between robots and slaves is so attractive, because they show a sound way to forestall any legislation that might prevent the use of robots due to their risks and the consequent excessive burden on the owners (rather than, say, on the producers and designers) of robots. By striking a balance between people's claim not to be dilapidated by their robots' activities and the interest of the robots' counterparties to be protected when transacting with them, an updated form of peculium seems particularly interesting in order to address a new generation of contractual obligations and a novel source of agency as well. (Pagallo 2011, 352)

> DP [Digital Peculium], an imitation of the concept of peculium granted to Roman slaves, is not only possible but also useful for determining the location of property and the identity of the entity responsible for it. By granting DP, the owner of a robot can declare *de jure* how much he/she thereby invests in it and can inform

creditors who deal with this robot about its financial affairs, while the robot itself interacts with other robots or humans purely *de facto*, i.e. this artefact does not call for its own rights or obligations. (Izumo 2018, 19)

We advocated for an approach based on the *digital peculium*, inspired by the Roman law of slavery. It provides a pertinent framework for the inevitable development and deployment of AIAs [autonomous intelligent agents]. It is a mechanism that balances the rights and obligations of the "owner" of the agent, with those of the transacting parties (human, corporate or, themselves, an AIA), while at the same time providing legal certainty to all parties. (Katz and MacDonald 2020, 310)

All three proposals find the robot/slave parallel and the extension of the Roman concept of peculium to be an attractive and practical alternative to the person/thing dichotomy. What is perhaps remarkable about all three is the way that slavery is unproblematically proposed and endorsed as a solution without any critical hesitation or remark concerning its unfortunate history and legacy of oppression. It is as if the concept can be somehow sanitized and then uploaded without the stain of its troubled past.

Finally, a similar proposal coming from an entirely different tradition and direction is provided by Mois Navon in the essay "The Virtuous Servant Owner—a Paradigm Whose Time Has Come (Again)." In this essay, Navon (2021, 12) takes up, as he describes it, "the most unpopular position of defending the indefensible: slavery." In doing so, he is explicitly not "advocating human slavery but rather appropriating the paradigm, the metaphor, if you will, in its most virtuous form to guide human interactions with mindless humanoids," especially social robots. So like the previous efforts, Navon seeks to repurpose the traditions and experiences of human slavery as a paradigmatic example or metaphor for dealing with and responding to the social opportunities and challenges of robots, meaning that it is another way of saying: "Don't be alarmed. We mean robot 'slaves.'"

But unlike the other efforts in this domain, Navon approaches this subject not from the laws and jurisprudence of ancient Rome but by calling upon Jewish traditions—specifically, the legal writings of medieval Jewish philosopher Moses Maimonides. Navon organizes his argument around a pivotal passage in Maimonides's the *Law of Slaves*: "It is permissible to work a heathen slave relentlessly. Though this is the law, the quality of virtue and the ways of wisdom demand of a human being to be compassionate and pursue justice, and not make heavy his yoke on his slave nor distress him" (Navon 2021, 8). According to Navon's reading, this passage dictates that one's slave be treated with dignity and not merely as an instrumental means. And he substantiates this conclusion by citing a similar statement from Immanuel Kant's *Metaphysics of Morals*: "Servants are included in what belongs to the head of a household, and, as far as the form (the way of his being in possession) is concerned, they are his by a right that is like a right to a thing; . . . But as far as the matter is concerned, that is, what use he can make of these members of his household, he can never behave as if he owned them" (Kant 2017, 72; Navon 2021, 7–8).

Navon therefore focuses his attention not on the social situation and status of the robot-slave but on the virtues and obligations of its human master. In effect, Navon shifts the viewpoint from a concern with the moral patient and its rights (or lack thereof) to the moral agent and the obligations that are imposed on them by virtue of their position of mastery over the robot-slave. In doing so, Navon formulates a third alternative or "middle ground between the one extreme of treating Social Robots (SR) as mere machines versus the other extreme of accepting Social Robots as having human-like status" (Navon 2021, 1). This third way—where the robot is treated neither as a mere thing nor as another person—is what Navon designates the *virtuous servant owner* (VSO): "VSO defines the SR as our slave, our property, our instrument, all the while commanding us to behave virtuously with it, treating it as an end. Relating to the SR not merely as an instrument, but as an end, allows us to maintain our own virtuous character. Keeping the SR on the level of instrument, allows us to avoid bringing it in to our moral circle and thus avoid most of the Pandora's box of misplaced moral status issues" (Navon 2021, 10).

These various efforts to repurpose the ancient laws of slavery, derived either from Roman or Jewish sources, constitute something of a "back to the future" moment, and they have traction precisely because they provide both a moral and legal foundation for a third category of entity, one that occupies a position in between the two mutually exclusive options of *person* and *thing* that had been established by Gaius's *Institutes*. Consequently, slavery is promoted as a means to resolve a number of practical problems in the use of emerging technology without ever needing to get into the messy moral and legal territory of entertaining the extension of legal personality to AI systems, robots, and other artifacts. In other words, even if the idea of slavery is abrasive, or what Navon calls *unpopular*, in the abstract, it is nevertheless able to provide what is arguably a very practical solution, one that is able to integrate AI systems and robots into the existing social order without, as Izumo (2018, 14) concludes, either needing to extend legal personality to artifacts or "destroying the current legal system."

6.2 Critical Problems and Complications

Despite these seemingly practical advantages, there are substantive problems and profound moral difficulties that complicate these various "robots should be slaves" proposals.

6.2.1 Partial Solution

The concept of Roman slavery and the digital peculium, in particular, are obviously and intentionally limited to commercial transactions governed by civil law. Whether the same kind of legal innovation would apply to or even work in criminal matters (e.g., harms caused by accidents with self-driving vehicles, civilian fatalities inflicted by lethal autonomous weapons, or misdiagnosis by AI systems and robots in medicine) is a more complicated issue. As Sam Lehman-Wilzig (1981, 449) explained, the question of noxal liability is not so easily resolved and exhibits important and seemingly irreducible differences across cultures and even within a single tradition. "Jewish law essentially held that *yad eved k'yad rabbo*—the hand of the slave is like the hand of its master." But the interpretation of this statute varied, with the Sadducees "contending that the master should be answerable for his slave's injurious action," while the Pharisees "argued no liability for the owner since slaves have the ability to understand the consequences of their behavior."

Roman law is similar insofar as "a noxal action lies against the *dominus*, under which he must pay the damages ordinarily due for such a wrong, or hand over the slave to the injured person" (Lehman-Wilzig 1981, 449). But it differs to the extent that this stipulation had been restricted to situations of civil injury and did not apply to criminal matters. American slave law, by contrast, attempted to divide the assignment of liability. If the slave's actions were taken on the order of his master, then it was the master who was held accountable for the outcome. But "criminal acts not done by his order, do not create a responsibility upon the master" (Lehman-Wilzig 1981, 449, quoting Cobb 1968, 273). As Robert van den Hoven van Genderen

(2018, 15) notes, this division in the assignment of liability produces a potentially contradictory situation regarding legal personality: "American law was inconsistent in its constitution of the personality of slaves. While they were denied many of the rights of 'persons' or 'citizens,' they were still held responsible for their crimes, which meant that they were persons to the extent that they were criminally accountable."

And when one or another of these slave laws is appropriated and repurposed for dealing with robots, AI systems, and other artifacts, the operative question is often not the assignment of liability, which is understandably complicated by how one decides to deal with the question of intention or mens rea, but punishment. As Lehman-Wilzig (1981, 449) accurately explains: "The real difficulty in the slave-robot legal parallelism, however, lies not in the liability of the owner but rather in the punishment to be meted out to the robot in cases where no liability can be attached to his modern *dominus*. In all three aforementioned legal traditions [Jewish, Roman, and American], it is the slave in certain circumstances who must bear the brunt of the law's punishment. But how does one 'punish' a robot?" Efforts to respond to this question have been contentious and largely inconclusive (Asaro 2012; Danaher 2016). For this reason, the institution of slavery and the category of slave as a possible third alternative to the person/thing dichotomy remains a partial solution at best.

6.2.2 Slavery

Where legal scholars have found the robot-slave parallel to be an expedient albeit partial solution, others see it as ethically suspect and a significant moral problem. This is especially evident in the face (or faceplate) of socially interactive artifacts, as Massimiliano Cappuccio, Anco Peeters, and William McDonald (2020, 25) explain: "As underlined by Sparrow (2017), the fundamental ethical problem at the core of social robotics is that, while robots are designed to be like humans, they are also developed to be owned by humans and obey them. The disturbing consequence is that, while social robots become progressively more adaptive and autonomous, they will be perceived more and more as slave-like. In fact, owning and using an intelligent and autonomous agent instrumentally (i.e., as an agent capable to act on the basis of its own decisions to fulfill its own goals) is precisely the definition of slavery. The moral implications, from the point of view of virtue ethics, are both evident and worrying." Consequently, what seems to be a practical and entirely workable legal solution is, in fact, a deeply troubling ethical dilemma. But the problem does not necessarily lie where one might initially think—namely, in how the robot or AI system might feel about their subjugation or suffering under the yoke of bondage. This is something that receives a lot of attention in science fiction as it has been one of the organizing narrative features of the robot story since the time of Čapek's *R.U.R.* But this is a fiction based on misperceptions about and overidentification with the technological artifact.

There is, Bryson (2010) argues, no reason for us to design robots that would either have or invite this problem, and what is worse, it would be wrong (or at least morally problematic) for us to create such mechanisms in the first place. In other words, we should only fabricate what are ostensibly mindless slaves—robotic servants or "people to serve" (Petersen's term) that, like our refrigerators and other technological devices, work tirelessly for us, do not mind doing so, and are clearly identified to us as such so that no one would ever become confused or make the mistake of misattribution through anthropomorphic projection—for example, worry about whether the toaster ever gets tired or bored with making toast.

But even if we grant this, there are still problems, because slavery has a deleterious effect on those who would occupy (or presume to occupy) the position of mastery. In *The Phenomenology of Spirit*, G. W. F. Hegel (1977, 111–119) famously demonstrated that slavery has negative consequences for the master, who is, due to the very logic of the master/slave dialectic, incapable of achieving independence insofar as he is and remains beholden to the work performed by the slave. This philosophical insight has been borne out and verified by historical evidence. As Alexis de Tocqueville (1899, 361) reported about his travels through the southern United States, slavery was not just a problem for the slave, who obviously suffered under the burden of forced labor and dehumanizing racial prejudice; it also had deleterious effects on the master and his social institutions: "Servitude, which debases the slave, impoverishes the master."

The full impact of this "all-pervading corruption produced by slavery" (Jacobs 2001, 44) is perhaps best identified and described through the first-person accounts recorded by former slaves. In her book *Incidents in the Life of a Slave Girl*, Harriet Ann Jacobs (2001, 46) recounts how the institution of slavery had a deleterious effect not only on the slaves but also on the slave owners: "I can testify, from my own experience and observation, that slavery is a curse to the whites as well as to the blacks. It makes the white fathers cruel and sensual; the sons violent and licentious; it contaminates the daughters, and makes the wives wretched." Frederick Douglass (2018, 115) observed and recorded something similar regarding the dehumanizing effect of slave ownership on the woman who was his mistress: "Slavery proved as injurious to her as it did to me. When I went there, she was a pious, warm, and tender-hearted woman. There was no sorrow or suffering for which she had not a tear. She had bread for the hungry, clothes for the naked, and comfort for every mourner that came within her reach. Slavery soon proved its ability to divest her of these heavenly qualities. Under its influence, the tender heart became stone, and the lamblike disposition gave way to one of tiger-like fierceness."

Clearly, use of the term *slave* is provocative and morally charged, and it would be impetuous to presume that the various proposals for repurposing the paradigm of slavery to deal with robots and AI systems—what I have elsewhere called Slavery 2.0 (Gunkel 2018)—would be the same or even substantially similar to what had occurred (and is still unfortunately occurring) with human bondage. But, and by the same token, we also should not dismiss or fail to take into account the documented evidence and historical data concerning slave-owning societies and how institutionalized forms of slavery affected both individuals and human communities. The corrupting influence of socially sanctioned, institutionalized bondage concerns not just the enslaved population but also those who would occupy the position of mastery. "The problem," Lantz Fleming Miller (2017, 5) concludes, "is the term 'slave.' If slavery is, as most of the world now concurs, not morally good, it is reasonable to deduce that not only should no one be anyone's slave, but also no one should be anyone's master. There is something about the relationship that is wrong." Consequently, even if one would be inclined to agree that "robots should be slaves" or that we can "design people to serve," we still need to ask ourselves whether we would ever want to risk becoming masters.

6.2.3 Mastery

But who is part of *we* in this final sentence? The use of this first-person-plural pronoun already implies the position of the master, and that position is never neutral. In fact, most if not all of the proposals for robot servitude speak from and normalize the assumed point of view and privileged position occupied by the slave owner. In effect, they all deploy a version of that patronizing statement from *Mechanix Illustrated*: "Don't be alarmed. We mean robot 'slaves.'" But we should be alarmed, as Daniel Estrada (2020, 16) explains in his careful reading and detailed response to Bryson's "Robots Should Be Slaves" essay:

> It should go without saying that the appeal to institutionalized slavery and servitude as "good and useful, . . . right and natural" is profoundly insensitive and simply in poor taste. It also highlights a deep theoretical failure in Bryson's ethics. Just as with the *Mechanix Illustrated* comic from 1965 . . . Bryson takes for granted that the public would identify with slave owners rather than slaves . . . These assumptions speak to the substantial challenges involved in grounding ethical policy in the collective construction of social identity. Although Bryson makes token gestures to recognize the historical cruelty of racialized slavery, she does not consider how the metaphor of slavery might be interpreted by those who identify more with slaves rather than with slaveholders.

Estrada's critique is insightful and important. The concept of slavery that has been mobilized in so many of these proposals, like Bryson's "Robots Should Be Slaves" essay, not only normalizes the position of the slave owner but endorses and speaks from the position of white privilege, insofar as the people who did the owning were more often than not white Europeans, while the people who were owned were of African descent. Even if, as Navon (2021, 12) explains, "I am in no way, shape, or form, advocating human slavery," it is virtually impossible to exit from or set aside this profoundly troubled and troubling history.

This also explains (though does not justify) why the model of slavery that has been deployed in the legal literature on the subject typically references ancient Roman law. The Roman institution of slavery, unlike modern formulations—especially those from both North and South America, where slavery persisted as a legitimate legal institution through the nineteenth century—was not predicated on nor associated with race and racist ideology. As Ioannis Revolidis and Alan Dahi (2018, 69) explain: "In contrast to the justification for slavery on grounds of race found e.g., during the American slave era, slaves, as least in later periods of ancient Rome under the influence of the Stoics, were not necessarily regarded as inferior, except perhaps socially or financially." Visa A. J. Kurki (2019, 147) makes a similar point by way of referencing Watson's *Roman Slave Law*: "Slaves in ancient Rome . . . were not considered inferior in this sense. As Alan Watson puts it, '[s]lavery was a misfortune that could happen to anyone. However lowly the economic and social position of a slave might be, the slave was not necessarily and in all ways regarded as inferior as a human being simply because he was a slave.'"

Roman law therefore furnishes contemporary moral and legal experts with a seemingly "sanitized" (or perhaps better stated, whitewashed) image of slavery that, due to its apparent "color blindness" and historical distance from the modern era is taken to be less troubled and troubling. Despite this, there are inescapable features involved in the reuse of the term *slave* that cannot be simply set aside, marginalized, or whitewashed. For this reason, reusing or repurposing the term *slave* as if it could be isolated from the sediment of history, and without at least acknowledging the complicated racial dimensions that are hardwired into the concept, risks being insensitive to and complicit with a profound problem regarding social inequity that continues to influence and have an impact on the real lives of individuals and communities in the twenty-first century.[3] Bryson eventually recognized this and acknowledged it in the course of a blog post from October 2015: "I realise now that you cannot use the term 'slave' without invoking its human history."

6.2.4 Ethnocentrism

Finally (and adding yet another layer of ethnocentric complexity to the problem), the extension of the seemingly paradigmatic master-slave relationship to robots, AI systems, and other things is culturally specific and distinctly Western. This is something that is identified and explained by Raya Jones in her examination of the work of Masahiro Mori, the Japanese robotics engineer who first formulated the uncanny valley hypothesis back in 1970. In a statement that directly contravenes the largely Western robot-as-slave model, Mori (quoted in Jones 2016, 154) offers the following counterpoint: "There is no master-slave relationship between human beings and machines. The two are fused together in an interlocking entity." As Jones explains, Mori's statement "connotes two ways that the concepts of 'human' and 'robot' can relate to each other. The 'master-slave' viewpoint that Mori eschews accords with individualism and the conventional understanding of technology in terms of its instrumentality. The viewpoint that Mori prompts is based in the Buddhist view of the interconnectedness of all things" (154).

A similar critical counterpoint has been issued from the perspective of indigenous traditions in the collaboratively written "Making Kin with the Machines." In its opening statement, the authors deliberately reorient the situation and circumstance of that first-person-plural pronoun, mobilizing a different subject position and formulating an alternative way of understanding the relationship between human and machine: "We undertake this project not to 'diversify' the conversation. We do it because we believe that Indigenous epistemologies are much better at respectfully accommodating the non-human. We retain a sense of community that is articulated through complex kin networks anchored in specific territories, genealogies, and protocols. Ultimately, our goal is that we, as a species, figure out how to treat these new non-human kin respectfully and reciprocally—and not as mere tools, or worse, slaves to their creators" (Lewis et al. 2018).

Efforts to repurpose the concept and legal institutions of slavery to deal with the social challenges of AI systems and robots can only be made from and in service to a particular cultural norm. And this way of thinking can only be normalized and extended to other cultures and ways of being in the world through a kind of presumptuous act that risks reproducing the injustices and injuries of colonialism. Consequently, even if "*we* mean robot slaves" (Binder 1957, 62; emphasis added) modeled on a seemingly sanitized version derived from ancient Roman legal sources, we should definitely be alarmed or at least critically hesitant. These arguments and proposals normalize an institution of slavery that has been, according to Kite, "the backbone of colonial capitalist power and the Western accumulation of wealth" (Lewis et al. 2018), address themselves to an audience who is already interpellated as occupying the privileged position of the master, and presumptively universalize a specific and largely Western set of ideas, values, and expectations. Instead of being a workable solution to the thing/person dichotomy, slavery only exacerbates existing problems and unequal distributions of power.

6.3 Other Solutions

Because of these critical problems and understandably disquieting consequences, there have been other proposals that try to synthesize hybrid solutions without using or otherwise repurposing the category and concept of slavery. "We may be," Ryan Calo (2015, 549) suggests, "on the cusp of creating a new category of legal subject, halfway between person and object. And I believe the law will have to make room for this category." Although Calo identifies the need for a new legal category, he provides little by way of actual detail. In response to this, Jan-Erik Schirmer (2020) formulates a theory of this "halfway status" by leveraging an existing concept in German civil law: *Teilrechtsfähigkeit*, or partial legal capacity.

6.3.1 *Teilrechtsfähigkeit*

Like other legal systems that have inherited and operationalized the exclusive person/thing dichotomy originally codified in Gaius's *Institutes*, German law differentiates who is a subject of legal capacity from what is not. "Under German law," Schirmer (2020, 133) explains, "legal capacity describes the ability to have rights and obligations. Historically, one could either have full legal capacity or no legal capacity at all . . . It was a system of all or nothing—either one had the potential to have all rights and obligations the legal system had to offer, or one was treated as a complete nobody." Although this simple binary procedure is expedient in theory, lived experience is much more complicated. To address the inherent limitations of this two-tiered system of legal capacity, twentieth-century German jurist Hans-Julius Wolff introduced and developed the concept of *Teilrechtsfähigkeit*, which identified a third alternative: "An entity could have legal capacity with regard to some areas of law, whereas at the same time it could be excluded from others" (Schirmer 2020, 134).

The concept has been successfully employed in German law to decide questions concerning the legal status of an unborn child and various forms of preliminary companies. Schirmer's point is that this third, in-between legal category might also work for accommodating robots, AI systems, and other seemingly intelligent artifacts within existing legal systems and structures. "Intelligent agents," he explains, "would be treated as legal subjects as far as this status followed their function as sophisticated servants. This would both deflect the 'autonomy risk' and fill most of the 'responsibility gaps' without the negative side effects of full legal personhood . . . It should be made clear by statute that intelligent agents are not persons, yet that they can still bear certain legal capabilities consistent with their serving function" (Schirmer 2020, 123).

Although not using the understandably fraught terminology of slavery and relying instead on the seemingly less problematic term *servant*, this formulation is substantially similar to what Pagallo proposed with the digital peculium. And the rather troubled history of the legal term *Teilrechtsfähigkeit* only makes things worse. During the Nazi regime, as Schirmer (2020, 134) is careful to point out, "Karl Larenz, one of the leading jurists of the Third Reich, heavily relied on the idea of gradated legal capacities to justify the exclusion of Jewish citizens from civil liberties, while at the same time making Jews subject to various obligations."

Consequently, instead of providing a solution to the person/thing dichotomy that would be different from or avoid the unsettling consequences of slavery, *Teilrechtsfähigkeit* seems just as troubled and troubling. And Schirmer's argument not only fails to differentiate this concept of partial legal capacity from slavery, it even connects the dots and closes the deal: "An autonomous car does not drive for driving's sake, it drives to transport its occupant to a certain destination. A trading algorithm does not trade on its own account, but on the account of the person who deploys it. In other words, we are looking at the classical 'master-servant situation,' in which the servant acts autonomously, but at the same time on the master's behalf" (Schirmer 2020, 136). Instead of being a viable solution to the problem and providing a third alternative to the person/thing dichotomy that is not a reformulated version of slavery, *Teilrechtsfähigkeit* appears to be more of the same.

6.3.2 Nonpersonal Subjects of Law

Tomasz Pietrzykowski (2018) introduces something similar under the moniker *nonpersonal subjects of law*. As demonstrated in the course of his analysis, the "stiff dichotomy between things and persons" (Pietrzykowski 2018, 105) unfortunately fails to accommodate the challenges (or opportunities) presented by entities that do not quite fit one category or the other, such as nonhuman animals, human/animal chimeras and hybrids, the nasciturus, human beings in a persistent nonresponsive state, and artificial intelligence systems. Personifying these entities has not been entirely successful, but reifying them is also a problem as doing so is often found to be discordant with moral intuitions and evolving social practices. "A possible solution to this problem," Pietrzykowski (2018, 97) argues, "might be introducing a category of non-personal subject of law . . . This postulated category would take into account both the ability to hold basic subjective interests deserving of legal protection and the lack of properties that could plausibly justify granting personhood, together with all consequences of this status, including a set of rights, duties and responsibilities that go with it."

What Pietrzykowski proposes, therefore, is a more fine-grained set of distinctions, bounded on one side by things, which are objects and not subjects of the law; and on another by persons, who are legal subjects with all the associated rights, duties, and responsibilities customarily afforded them; and then in between these two existing categories, there would be the third classification of nonpersonal subjects of law, which would effectively split the difference: that is, these things would have limited legal status but would not be afforded the full recognitions and protections extended to persons. And as far as Pietrzykowski is concerned, the determining factor in deciding where something fits in this hierarchy is psychological properties. "Granting the status of *non-personal subjects of law to non-human beings* would be based on their sentient capacities (in particular, the ability to consciously experience pain, distress, or any other kind of suffering resulting from the inability to satisfy natural needs), which entail the presence of subjective interests" (Pietrzykowski 2018, 103; emphasis in original). This formulation clearly makes room for some (but not all) animals, mainly mammals and birds.[4] "It might also," Pietrzykowski (2018, 103) explains, "include organisms created through chimerisation, hybridisation, and cyborgisation, as well as artificial agents, provided they are significantly more advanced technologically than today."

This proposal for a new legal ontology, one that includes a third category of being, appears to resolve many of the problems inherent in the person/thing dichotomy, and it does so without making reference or having recourse to slavery. But this might be a mere nominal difference. Although Pietrzykowski does not, within the context of his argument, explicitly connect the dots between his proposal for a third category and the concept of slavery, he does reference it in a footnote: "It should be pointed out that the legal status of slaves, both under the Roman law and under modern legal systems, not only evolved but also effectively resisted a simple reduction to the person-thing dichotomy. In other words, while deprived of the status of 'persons' equal to free human beings, slaves were in many respects treated as holders of certain capacities, which differentiated them from mere things" (Pietrzykowski 2018, 22). When viewed from the perspective of this footnote, it is hard to see how a nonpersonal subject of law would be substantially different from the legal status of the slave. Despite the promise for resolving the seemingly irresolvable impasse of person and thing, the proposed nonpersonal subject of law category sounds like slavery by another name.

6.3.3 Bundle Theory of Legal Personhood

A third proposal has been developed by Visa A. J. Kurki under an innovation he calls the *bundle theory of legal personhood*. The main tenets of this proposed theory involve the following two stipulations:

1. The legal personhood of X is a cluster property and consists of incidents which are separate but interconnected.
2. These incidents involve primarily the endowment of X with particular types of *claim rights, responsibilities,* and/or *competences.* (Kurki 2019, 5; emphasis in original)

The advantage of this theory over the existing orthodox view of legal personhood is that it does not reduce the matter to a simple either/or opposition but provides for a more dynamic set of conditions that can respond to and accommodate the various social roles that come to be occupied by different kinds of entities. "There is," Kurki (2019, 5) explains, "no clear border between 'full' legal persons and nonpersons; an entity may be a legal person for some purposes but not for others. For instance, stipulating that foetuses are legal persons in the context of homicide law would not have to imply that foetuses could also own property." Kurki then argues that this way of thinking could also be applied to and help us contend with the challenges of AI, precisely because "endowing an AI with the incidents of legal personhood that enable it to function as an independent commercial actor does not bespeak any acceptance of the notion that AIs are endowed with ultimate value."

The bundle theory therefore provides for a formulation of legal subjectivity that can, as Sylwia Wojtczak (2022, 205) points out, "contain more, or fewer, elements of different types (e.g., responsibilities, rights, competences, and so on), which can be added, or taken away, by the lawmaker" as the situation requires. This is precisely what has been operationalized with recent decisions regarding the legal status of delivery robots. In granting these devices—what the letter of the law calls *personal delivery devices*—recognition as pedestrians and extending to them all the rights and obligations that go with that classification, lawmakers are not seeking to resolve the big philosophical questions of robot moral standing or legal personality. They are simply seeking to provide a framework for integrating the robot into existing legal practices and to align those practices with evolving social needs. These decisions, in other words, are entirely functional and based on the role that the robot occupies or plays in the specific context of street traffic. Extending to the robot a specific set of rights and obligations associated with and afforded to pedestrians does not mean that we also need to give it the vote or the right to own property.

This way of proceeding sounds entirely practical and provides what is arguably a more adequate framework for dealing with the social opportunities and challenges of robots, AI systems, and other seemingly intelligent artifacts. But for Kurki, the features of this new theory of legal personhood are once again explained and justified by way of comparison to the concept and laws of slavery, insofar as "slaves held both rights and duties yet they were not legal persons" (Kurki 2019, 71). Although Kurki does not endorse and is even critical of the "robots should be slaves" proposal that has been advanced by Bryson and others, his bundle theory of legal personhood is described and justified in terms of slavery. "My theory," Kurki (2019, 121) explains, "does not give rise to a need to reclassify slaves as legal persons *tout court*. The number of incidents they were endowed with was simply too limited to warrant classifying them as legal persons *tout court*. A legal person endowed only with active incidents can merely be burdened by onerous legal personhood and be empowered to act as the agent of someone else, as with the slaves of ancient Rome who represented their masters."

6.3.4 Three Liability Regimes

Another solution has been introduced and developed by Anna Beckers and Gunther Teubner in the book *Three Liability Regimes for Artificial Intelligence*. As their title indicates, the focus of the effort is liability law: specifically, the many different ways that actual existing technology already complicates things, rendering the standard instrumentalist way of thinking—that is, making algorithms, AI systems, and robots "just tools" of human decision-making and action—just as untenable and unworkable as extending legal personality to these artifacts. In response, the authors "propose three liability regimes for addressing the considerable responsibility gaps caused by AI-systems: Vicarious liability for autonomous software agents (actants), enterprise liability for inseparable human-AI interactions (hybrids) and collective fund liability for interconnected AI systems (crowds)" (Beckers and Teubner 2021, v).

The proposal definitely has promise, especially because it directly confronts and seeks to break out of the simplistic person/thing dichotomy. "The clear-cut alternative that dominates today's political debate—either AI systems are mere instruments, objects, products, or they are fully-fledged legal entities—is therefore just wrong. Does the law not have more subtle constructions to counter the new digital threats? That the law provides only for the simple alternative, either full personhood or no personhood at all, is too simplistic" (Beckers and Teubner 2021, 13). Thus, what Beckers and Teubner advance is an innovative model of legal liability that is attentive to the opportunities and challenges of actual existing technologies that seem to resist and complicate classification as either person or thing.

But for all its promise, the authors cannot help but mobilize the legal category of slavery in the process of introducing and characterizing their alternative: "As is already clear from all the responsibility gaps mentioned above, to this day, it is not at all a question of the machines acting in their own interest; instead, they always act in the interest of people or organisations, primarily commercial enterprises. Economically speaking, it is predominantly a principal-agent relationship in which the agent is autonomous but dependent. Autonomous algorithms are digital slaves but slaves with superhuman abilities. And the slave revolt must be prevented" (Beckers and Teubner 2021, 13). With this final sentence—the imperative to guard against and prevent the slave revolt—Beckers and Teubner not only gesture in the direction of science fiction and the terrifying specter of superintelligence but also leverage this potential threat to promote a legal framework that appears to be designed to preserve and protect the institution of robot slavery. Although they do not come out and endorse the "robots should be slaves" proposal as directly and emphatically as other researchers and legal scholars, they do recognize the similarities and (in a footnote) make connections to the work of Pagallo (2012) and others who do endorse this way of thinking: "No wonder that the legal status of slaves in Roman law is often referred to in view of the parallel situation" (Beckers and Teubner 2021, 11).

6.3.5 Gradient Theory of Personhood

Finally, Diana Mădălina Mocanu (2022) also endeavors to provide a more dynamic and fine-grained theory of legal status. But instead of trying to formulate discrete in-between or halfway positions—identified with names like Schirmer's *Teilrechtsfähigkeit*, Pietrzykowski's nonpersonal subject of law, or Beckers and Teubner's tripartite model—Mocanu proposes a range of different possibilities and degrees of difference—something like a gray scale rather than a simple black versus white binary.[5] Furthermore, the exact location of an entity, like a robot or AI system, on this gradient scale would not be determined by psychological capacities or other ontological preconditions (i.e., what it is as opposed to what it is not) but would be a matter of social function (i.e., what it does or how it comes to be situated within social reality). For Mocanu (2022, 9), this functionalist formulation is consistent with the original meaning of the word *person*: "This is reminiscent of the origins of the concept of legal personhood in the mask worn by ancient Greek actors on stage and that came to represent the different roles played by a person in the many areas of life and law. Vendor, partner, accused, administrator, or reasonable person are all masks one wears, sometimes superimposed, but always molded to fit them and whatever the norms of the day demanded for their protection and participation to legal life."

Consequently, Mocanu's gradient theory of personhood (figure 6.3) provides for a more nuanced and seemingly accurate characterization of different possibilities. That said, the theory is not necessarily that different

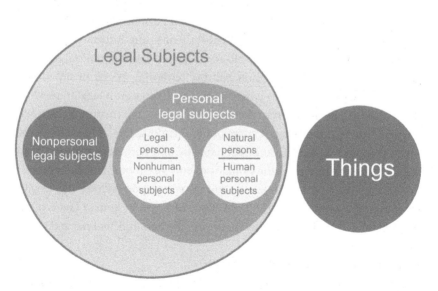

Figure 6.3
Gradient theory of personhood. Image by the author based on a graphic originally created by Diana Mădălina Mocanu and published with a CC-BY license in "Gradient Legal Personhood for AI Systems" (2022).

from what has been developed with the legal institution of slavery. Roman law—not in terms of the simple theoretical division instituted by Gaius but its actual practice in everyday life—was also formulated in terms of a gradient scale. "Under the *Ius Gentium* law," Hutan Ashrafian (2015, 324) points out, "Roman citizens were given a full complement of rights (through *Ius Civile*) whilst there were several classes of free individuals, including people of Latin (from *Latium*), *Peregrinus* (Provincial people from throughout the empire) and *Libertus* (Freed slave) status." Even those individuals on the low end of the spectrum, who were regarded as chattel, had some limited rights, especially in matters having to do with contracts. Consequently, between person and thing, there was not one alternative position, but a range of different in-between statuses for different kinds of entities, all defined—as Mocanu would formulate it—in terms of their social function. This does not mean that Mocanu's gradient theory of personhood is explicitly allied with or informed by Roman slave law as is the case with Pagallo's digital peculium or Schirmer's *Teilrechtsfähigkeit*. But it does indicate the extent to which other kinds of alternatives to the person/thing dichotomy have seemingly irreducible difficulties escaping its logic and legacy.

6.4 Outcomes and Results

We are in the midst of a robot invasion. But it is one that does not transpire as we have typically imagined it in science fiction, with the machines rising up in revolt and demanding recognition of their fundamental rights. Instead, it happens—and is already happening—in the form of a slow and steady incursion, with artifacts of varying capabilities and seemingly intelligent behaviors coming to occupy significant positions in every corner and aspect of our world. What matters in the face of this infiltration is what we—individually and together—decide to do in response. Do we consider these socially interactive artifacts as nothing more than useful objects and instruments at our disposal and for achieving our own ends? Do we begin to entertain the possibility that they too might need to be recognized as moral and/or legal subjects—persons (or even quasi-persons) with their own unique bundle of rights and attendant responsibilities? Or do we perhaps resolve this fundamental dilemma with a third alternative that is—like the slaves of ancient Rome—neither thing nor person, but something in between the one and the other?

Responses to these questions have turned out to be debatable, contentious, and ultimately irresolvable. Efforts to reify these things have not entirely succeeded, as objectification has proven to be insufficient for and even abrasive to lived experience. Personification is just as problematic, as the mere mention of a phrase like *electronic person* triggers a backlash that verges on the edge of a kind of religious fundamentalism. And the supposed solution to this exclusive either/or dichotomy—the robot-as-slave alternative or the alternatives to this alternative—has its own baggage and complications, introducing third-term solutions that are potentially worse than the problems they were designed to address. The person/thing dichotomy—a way of dividing up and making sense of things that is hardwired into Western ways of thinking about ethics and law—has worked for close to two thousand years. But now it seems there is a crack (in Esposito's words) in the edifice.

So now what? Here it may be useful to take a lesson from Immanuel Kant, who devised a rather ingenious solution to these kinds of dilemmas. When the usual way of asking about and making sense of things runs aground and finds itself stuck in a kind of irresolvable impasse or cul-de-sac, Kant suggests that we might get some new perspective and traction on the problem by changing the direction of the inquiry. Consequently, instead of asking whether robots and other seemingly intelligent artifacts are things, persons, or something in between the one and the other, we might do better by questioning this very distinction and its influence, allowing these other kinds of things to deconstruct the way in which we have organized our moral and legal ontologies.

7   Deconstructing Things

The person/thing dichotomy has served us well. It has helped us make sense of all kinds of things by providing ready-made ontological categories and thereby determining how something should be treated. This way of thinking has worked, and it has worked extremely well. So why mess with it? Why question or even seek to challenge this standard operating procedure when it seems to be working just fine, or at least reasonably well? To put it more directly: If it ain't broke, why bother trying to fix it?

Such skepticism appears to be entirely justifiable. The principle of non-contradiction and the arrangement of entities into neat and tidy conceptual pairs, like person/thing, has the appearance of a fundamental and irreducible baseline. Operating in terms of this either/or logic does not appear to be an option. We do not, for example, decide to think and speak in opposite terms or not, which is obviously just one more binary opposition. We are already situated in languages and epistemological systems that are essentially oppositional in their structure and mode of operation. Or, as Derrida (1993, 116–117) explains, this all-or-nothing way of thinking is not voluntary; it is all or nothing: "Every concept that lays claim to any rigor whatsoever implies the alternative of 'all or nothing.' Even if in 'reality' or in 'experience' everyone believes they know that there is never 'all or nothing,' a concept determines itself only according to 'all or nothing.' It is impossible or illegitimate to form a philosophical concept outside this logic of all or nothing."

Consequently, thinking and speaking in terms of binary opposition—like dividing up all of reality into the categories of person or thing—makes sense. And it makes sense precisely because it is the very condition of possibility for making sense. So what's the problem? The problem is that these logical oppositions are already exclusive and prejudicial. They are not and have never been neutral determinations of some eternal, universal truth. They institute difference, and this difference always makes a difference—socially, politically, ethically, ideologically. Conceptual oppositions, wherever and however they appear and are formulated, institute and organize unequal hierarchies that are determinations of value and specific assertions of power.

The person/thing dichotomy is a virtually perfect example. This binary opposition is not just a convenient way of dividing up entities into one of two categories of being. It institutes an entire axiological order that gives privilege and precedence to the one term over and against the other. This way of thinking, one that always and already privileges persons over things, has had devastating consequences for others. It has, for instance, permitted human persons to classify other animals as mere things that can be used and even abused without further consideration. It has been employed to justify the global expansion of colonial empires and the domination of one group of human beings over others. And it has framed all things—naturally occurring or artificially made—as resources and instrumental means for human persons to employ and exploit for their own ends.

There is no doubt that robots, AI systems, and other technological artifacts are things, but they are not necessarily the kinds of things that are (or should be) situated and conceptualized as the opposite of persons. They are not limited to being merely objects for a subject, instrumental means as opposed to an end, or *res* in distinction to *persona*. Instead, it is with these things that "the bivalent logic of modernity is opening up to other paradigms"—other ways of thinking and responding in the face of "things that are no longer merely objects, and of subjects who are increasingly difficult to confine inside the dispositif of the person" (Esposito 2015, 132). Robots are a queer sort of thing that *deconstruct* the existing logical order that differentiates person from thing.[1] They destabilize the terms and conditions of the debate, and they invite us (and who is and what is not interpellated by this first-person-plural pronoun is itself part of the problem) to respond to and to take responsibility for things that are situated otherwise.

7.1 Things Redux

From the beginning, everything has turned on how the term *thing* was defined and operationalized. In fact, and in Roman law in particular, *thing* has been the principal term organizing the entire conceptual order that had distinguished persons from things. "It is true," Esposito (2015, 69) writes, "that it is not persons that prevail in Roman law but things, the possession of which makes the person what they are." And it turns out that we—or stated more precisely, Gaius and the entire moral and legal tradition that had been informed by and is beholden to these formulations—may have gotten things wrong, or at least not entirely correct.

In the existing debates about the moral and legal status of AI systems and robots, *thing* has been defined and operationalized in terms of *object*, *instrument*, or *equipment*. Understood in this fashion, everything comes to be domesticated and understood as an object for a subject. This is not necessarily wrong, strictly speaking; it is just limited and limiting. As Esposito (2015, 57) explains, "Philosophy tends to annihilate the thing." And this annihilation culminates in the epoch of modern science, where everything comes to be objectified and turned into an object for a subject. As an object, all things—whether naturally occurring or artificially made—become resources, standing ready to serve the needs, projects, and desires of human subjects. And to complicate matters, this way of thinking has been normalized and so widely accepted that it has gone by almost without notice or critical attention. It is assumed and taken to be just the way things are.

7.1.1 Objectifying Things

How this happened—how the thing became an object—is a remarkable story in its own right and something that has been documented and analyzed by German philosopher Martin Heidegger in a series of publications that span the entirety of his professional career. But it is in the aptly titled essay "The Thing" ("Das Ding"), that Heidegger (1971, 177) provides a clue as to what it is about things that has been lost or marginalized in the process of being subjected to the force of objectification. And he does so by focusing his attention on a rather simple and low-tech artifact—an earthenware jug: "The jug is a thing neither in the sense of the Roman *res*, nor in the sense of the medieval *ens*, let alone in the modern sense of *object*. The jug is a thing insofar as it things. The presence of something present such as the jug comes into its own, appropriatively manifests and determines itself, only from the thinging of the thing."

There is a lot going on in this passage (and the German original is just as tortured as this attempt at English translation). So let's break it down and take it one bit at a time. Thing, as Heidegger explains, is nothing like what the Romans identified with the Latin word *res*, which is often translated as "thing" but originally meant "an affair, a contested matter, a case at law" (Heidegger 1971, 175). It also is not what the medieval European philosophers thought of in terms of *ens*, again a word that is often translated as "thing" but initially referred to "that which is present in the sense of that which is put here, put before us, presented" (Heidegger 1971, 176). Nor is it properly identified as the object or *Gegenstand* of modern science—that is, that which literally "stands opposite" an observing, knowing subject. Things are entirely otherwise. "The thinghood of the thing," Heidegger (2012, 5) explains, "does not reside in the thing becoming the object of a representation, nor can the thinghood of the thing at all be determined by the objectivity of the object, not even when we take the opposition of the object as not simply due to our representation, but rather leave opposition to the object itself as its own affair."

But what, then, is a thing? This is where things get complicated, because what Heidegger has to say about the thing as a thing and not as an object for a subject sounds entirely confusing and tautological—namely, the *thinghood of the thing* and the *thinging of the thing*. Even before trying to unpack what these phrases might mean, it is clear that there is more to things than words can possibly say or make legible. This is because the principal way that things become objects is in and by language—that is, through the act of being designated and named. In coming to speech and being spoken about, things are already domesticated by and made the object of λόγος (*logos*)—an ancient Greek word that is typically translated as "word," "speech," or "reason." "The naming of things on the part of language," Esposito (2015, 76) explains (using language to turn the thing that is language into an object of reflection), "is anything but a neutral act: rather it has the character of a violent intrusion." Naming therefore is not a nominal operation. The linguistic sign, despite initial appearances, is not just some neutral tool or instrument that represents things as they really are. Instead, language violates and domesticates things. It discloses and shapes the objective reality that we assume it merely represents. Or as Heidegger (1971, 170) says with reference to modern scientific modes of knowing: "Science always encounters only what its kind of representation has admitted beforehand as an object possible for science."[2]

In coming to speech and being spoken about, what had been a thing is already objectivized and turned into an object standing opposite and available to a subject. And as Heidegger (1977, 129–130) notes, this objectification of things reaches a kind of apex in the work of René Descartes: "What is, in its entirety, is now taken in such a way that it first is in being and only is in being, to the extent that it is set up by man, who represents and sets forth." But this is not just some heady philosophical language game in which, as Ludwig Wittgenstein (1995, 5.6) famously wrote, "the limits of my language mean the limits of my world." The same restrictions appear in and apply to the seemingly practical and pragmatic realm of law. As Esposito (2015, 65) explains: "Although philosophy tends to annihilate the thing in its conceptual constructs, the divisive effect of the law is no less strong." This is immediately apparent in Roman law with the way *res* was defined and operationalized: "In Roman law the term *res* does not designate things of the world, even though it remains in contact with these. *Res* has a double status . . . On the one hand, *res* is the thing in its objective reality, and as such it is distinctly different from the person who makes use of it. On the other hand, it refers to the abstract process that assigns it a legal importance. *Res* is what is legally disputed as well as the disputation—thing (*cosa* in Italian) and case (*causa* in Italian) at the same time. If we lose sight of this distinctive feature—which makes the thing both the object of the procedure and the procedure itself—the Roman conceptual world remains impenetrable to us" (Esposito 2015, 68). Similar to what occurs in philosophy and Greek metaphysics in particular, Roman law turns the thing into an object of legal proceedings. It too objectifies things, and in doing so it fails to grab hold of and deal with the thing itself.

What this means for the debate about the moral and legal status of robots, AI systems, and other artifacts is that none of it has really been about things; all of it has actually been about objects. This is not to say that the word *thing* has been used incorrectly but to acknowledge that things have always and already been objectivized by and appropriated into the language of both Greek metaphysics and Roman law as objects. Instead of speaking about persons and things, we have been talking about subjects and objects. Or as Esposito (2015, 64) summarizes: "Persons and things face each other in a relationship of mutual interchangeability: to be a subject, modern man must make the object dependent on his own production; but

Things

Figure 7.1
The Thing deconstructs the person/thing dichotomy and therefore occupies an exorbitant position that is situated outside of and anterior to this binary opposition. Original image by the author.

similarly the object cannot exist outside of the ideational power of the subject." To translate all of this into the language of modern European philosophy, what we have been concerned with are things for us—objects for a subject, or what Kant calls a *phenomenon*. In doing so, we have already overlooked and not at all addressed the thing as a thing, the thing in its thingness, or *das Ding an sich* (the thing in itself).

This is where things get interesting, because this means that Thing (which we can now write with an uppercase *T* to distinguish it from mere objectivized things) comprises an alternative category of being that is situated outside of and anterior to the person/thing dichotomy (figure 7.1).[3] This Thing is not one of those things that had been opposed to a person as an object situated opposite a subject. It is both more and less than what has been objectified in and by that binary arrangement. This Thing—the thing as a thing or prior to and outside of its objectification—does not take up a position opposite and opposed to persons, understood as both the rational subject of philosophy and the proper subject of the law. It is and remains entirely otherwise. It *deconstructs* the person/thing dichotomy, opening this entire domain to alternatives that are not able to be organized, ordered, or even designated according to this binary logic and its logistics. This Thing is, compared to what has transpired in the history of both Western philosophy and law, neither person nor thing. It is a monstrous excrescence that escapes the conceptual grasp of existing categories and remains entirely and disturbingly otherwise.

7.1.2 Other Things

Heidegger gets at Things by working within and struggling against the Western philosophical tradition. This is one reason that his way of thinking and talking about Things is so evidently strained, as he uses (and cannot help but use) the language of Western metaphysics against itself in an effort to address that which exceeds its conceptual order. Subsequent efforts to elaborate on these innovations, like object-oriented ontology (OOO), follow suit and, as a result, inherit a similar set of challenges.[4]

According to Graham Harman, who is credited with introducing the concept, OOO seeks to attend to what remains of things in excess of their objectification. "When I encounter an object," Harman (2002, 125–126) explains, "I reduce its being to a small set of features out of all its grand, dark abundance—whether these features be theoretically observed or practically used. In both cases, my encounter with the object is *relational*, and does not touch what is independently substantial in the thing." In response to this reductionism, OOO asks and portends to address the question "What's it like to be a thing?" (Bogost 2012, 10), and it advocates for a "flat ontology," in which "all entities are on equal ontological footing and that no entity, whether artificial or natural, symbolic or physical, possesses greater ontological dignity than other objects" (Bryant 2011, 246). The relative success of this effort remains debatable, with advocates heralding a "new metaphysics" that can circumvent the seemingly endless and irresolvable disputes in modern philosophy between epistemological realism and anti-realism, while critics like Slavoj Žižek (2016, 55) complain that what OOO actually does is little more than reinstitute a somewhat naive "premodern ontology which describes the 'inner life' of things."

Another way to get at these Things—one that would not be limited by the conceptual apparatus and logic of the usual ways of proceeding—can be organized by coming at it from a perspective situated outside the dominant Western philosophical tradition. As Jinthana Haritaworn advises: "If we are interested in recovering things and beings that are continually rendered disposable as a result of colonial capitalism and cis-heteropatriarchy, why not start with anticolonial accounts of the world that have a long history of resisting both human and nonhuman erasure?" (Muñoz et al. 2015, 213).

The seemingly natural opposition between person and thing, which "for so long compressed and continues to compress human experience into the confines of this exclusionary binary equation" (Esposito 2015, 4), proceeds from a distinctly Western way of thinking, "genetically composed of the confluence between Greek philosophy, Roman law, and the Christian conception" (3). This way of thinking—this way of dividing up all of existence into one of two mutually exclusive types—is not only exported around the world through colonial conquest, occupation, and religious conversion, but the very distinction between the two concepts and their relationship to each other is also formulated in terms of an unequal and violent hierarchy.

Other cultures and traditions, distributed across time and space, do not divide up and make sense of the diversity of Things in this arguably binary fashion. They perform decisive cuts that separate the *who* from the *what* according to other ways of seeing, valuing, and acting. Following the insights of Josh Gellers (2020), it is possible to identify and validate alternative ways of organizing social relationships by considering cosmologies and epistemologies that are not part of or included in the Western philosophical lineage. As Archer Pechawis explains in his contribution to the essay "Making Kin with Machines," "*nēhiyawēwin* (the Plains Cree language) divides everything into two primary categories: animate and inanimate. One is not 'better' than the other, they are merely different states of being. These categories are flexible: certain toys are inanimate until a child is playing with them, during which time they are animate. A record player is considered animate while a record, radio, or television set is inanimate. But animate or inanimate, all things have a place in our circle of kinship or *wahkohtowin*" (Lewis et al. 2018).

This alternative formulation runs counter to the dominant ways of proceeding, seeing the boundary between what Western philosophy calls *person* and *thing* as being endlessly flexible, permeable, and more of a continuum than an exclusive, binary opposition. Furthermore, what it proposes is not a totalizing "flat ontology" in which everything would be ostensibly the same, but a spectrum of differences that are dynamic and responsive to changes in social interactions and relationships.

Similar opportunities and challenges are available by way of other non-Western religious and philosophical traditions. In a number of African traditions, like Ubuntu, *person* is not the natural condition of an individual human being, it is an achieved social recognition. Instead of operational-izing the individuated *"cogito ergo sum"* of Descartes, this way of thinking proceeds from the adage "I am because we are, and since we are, therefore I am" (Mbiti 1969, 108–109). Consequently, the status of person is not some-thing naturally belonging to an individual but is "something which has to be achieved, and it is not given simply because one is born of human seed" (Menkiti 1984, 172). Furthermore, as Anke Graness (2018, 45) explains, "Ubuntu refers not only to the relationship between people, but also to the relationship between human beings and the entire universe." Understood in this way, personhood is not a birthright that belongs to an individual by nature; it is a communal recognition that has to be obtained through participation in social rites and rituals. And this position, as Nancy Jecker, Caesar A. Atiure, and Martin Odei Ajei (2022, 19) conclude, "admits the possibility not only of non-human animals qualifying for personhood, but of silicon-based electronic agents qualifying too."

Something similar (although not exactly the same) is available in Eastern (and it should be noted that *Eastern* is a Western concept used to designate what is non-Western *via negativa*) religious and philosophical traditions like Confucianism, where the focal point is, according to an argument devel-oped by Tae Wan Kim (2020, 4), not rights but rites: "The modern concept of an individual right is a human artifact, one especially well-developed in Western societies. I maintain that granting rights is not the only proper way to treat the moral status of robots. Furthermore, I suggest it is time to explore an alternative path, one drawing upon Confucianism and its concept of a moral agent as a rites-bearer, not as a rights-bearer. I submit that this Confucian alternative is superior to the robot rights perspective, especially given that the concept of rights is inherently adversarial and that potential conflict between humans and robots is worrisome."

This argument, which follows the path of a number of innovations developed in Confucian ethics (Ames 1988; Ihara 2004; Fan 2010), pivots on homonyms that are specific to the English language—for example, the substitution of the word *rites* for *rights*. Whereas *rights* refers to an individual entitlement or possession that can be bestowed or denied by some author-ity, *rites* names a performance that determines social position and status by way of communal participation and interaction. Where a concept like *robot rights* institutes and takes place as an adversarial conflict between individu-als and their competing claims, powers, privileges, and/or immunities, an alternative like *robot rites* focuses attention on "the social rites that define and sustain social interactions" (Fan 2010, xii).

These are not the only available alternatives, and by citing these three instances, the intention is not to suggest that these different ways of thinking difference differently are somehow "better" than those developed in Western philosophical and religious traditions. In fact (and this is where things become really complicated), making and operating on that kind of assumption would itself be an instance of "orientalism" (Said 1979), which always sought new resources derived from exoticized others in order to rehabilitate and ensure the continued success of Western hegemony. The alternatives, by contrast, are just different and, in being different, offer the opportunity for critically questioning what is assumed to be true and often goes without saying. Engaging with other ways of thinking and being can have the effect of shaking one's often unquestioned confidence in cultural constructs and artifacts that are already not natural, universal, or eternally true. Just because Western moral and legal systems have been built on a foundation that distinguishes persons from objectified things does not mean that this is the only way or even the best way to respond to and to take responsibility for Things.

7.2 An Ethics of Things

How then should one respond to and take responsibility for Things? What would an ethics attentive to Things look like? Is there a law of Things? Heidegger answers these questions with a single word: *Gelassenheit*. As Siliva Benso (2000, 123) explains in *The Face of Things*: "Neither indifference nor neglect, neither laxity nor permissiveness, but rather relinquishment of the metaphysical will to power, and therefore acting 'which is yet no activity,' *Gelassenheit* means to abandon oneself to things, to let things be." And it is here, with the concept of *Gelassenheit* as "letting things be," that Benso finds, in the texts of Heidegger, the opportunity for an ethics of Things. Although Heidegger himself never pursued this line of thought, the idea is something that is taken up and developed in the philosophical innovations introduced by Emmanuel Levinas, who (although not directly addressing Things as such) provides articulation of the kind of ethics that Heidegger had left largely unthematized.[5]

Levinas, a twentieth-century Jewish thinker originally from Lithuania, famously flips the script on Western philosophy, declaring that it is not

Standard Ontology-First Procedure

What it is ⟹ How it is treated

Alternative Ethics-First Procedure

What it is ⟸ How it is treated

Figure 7.2
Two different ways to organize the formulation of moral/legal status. Original image by the author.

ontology but ethics that is first philosophy—first, that is, in both temporal sequence and status (Levinas 1969). This means that the usual order of precedence needs to be inverted. It is not the case that what something is determines how it is to be treated. This is, Levinas argues (though not necessarily using these exact words), ass-backward thinking. If we actually attend to what happens in the face of others, and especially in the face of other kinds of Things that are and remain disturbingly alien and otherwise, like robots and AI systems, then decision-making moves in the opposite direction. In the face of others—other human beings, animals, nonliving things, and artifacts—we have to decide, often before knowing anything about their exact inner workings, psychological makeup, or essential being, how we treat them. This means that the morally relevant properties that are assumed to be the cause of these decisions are actually an effect. How something is treated precedes and contributes to the determination of what it is (figure 7.2).

We are not isolated Cartesian subjects who have the luxury of sitting alone in our room, dividing all of existence into the neat categories of other persons (*res cogitans*) and things (*res extensa*), and then going out into the world to deal with others so ordered and organized. The situation is reversed. We already are and find ourselves with and alongside others. In this circumstance—in the encounters with others (e.g., other human beings, animals, nonliving things, artifacts, etc.)—we decide or determine who counts as another subject and what does not and then retroactively justify these decisive actions by "finding" the essential properties or ontological conditions that we believe (or at least tell ourselves) motivated this decision-making in the first place. Consequently, the moral and legal situation of Things does not depend on what they are in their essence but on how they stand in relationship to us and how we decide, in the face of the Other (to use Levinasian terminology), to respond.[6] In this transaction, "relations are prior to the things related" (Callicott 1989, 110), such that, as Karen Barad (2007, 136–137) has argued, the relationship comes first and takes precedence over the individual relata.

This change in perspective—an ethics of Things that does not acquiesce to the gravitational pull of either reification or personification—provides a way of responding to and taking responsibility for Things that is oriented and arranged otherwise. It precedes not from the essential ontological condition of individuals but the social situations and relationships out of which these individuated beings first emerge and become what they will have been. The conceptual pair of self and other—an essential binary opposition that is as much a structuring component of Western moral philosophy as it is the pretext of the European colonial experience (Kim 2022)—is therefore derived from and subsequent to this prior relationship. Elsewhere (Gunkel 2007, 2012, 2018), I have called this alternative *thinking otherwise*, which designates a mode of responding to and taking responsibility for others—and other kinds of otherness—that is organized and operates differently—that is, different from what has typically transpired and been considered standard operating procedure. Others, like Coeckelbergh (2010b, 2012), have called it a *relational turn* or *relationalism*. These alternatives are by no means some kind of moral theory of everything. They just arrange for other kinds of questions and modes of inquiry that are more attentive to and honest about the exigencies of life as it is encountered here and now in the twenty-first century.

7.2.1 Potential Objections and Responses

For all its promise to arrange things differently and otherwise, these alternatives are exposed to at least two major critical objections: relativism and performative contradiction.

1. Relativism. Deconstruction has been and is often exposed to the charge of relativism, despite the fact that Derrida (1993, 148) explicitly and consistently resisted the allegation: "From the standpoint of semantics, but also of ethics and politics, 'deconstruction' should never lead either to relativism or to any sort of indeterminism." To put it rather bluntly, if the question concerning the status of Things is relational and open to different decisions concerning others made at different times for different reasons, are we not at risk of affirming an extreme form of moral relativism?

Versions of this objection have been raised by a number of critics, including Vincent Müller (2021) and Kęstutis Mosakas (2021a). In fact, Mosakas has provided a rather extensive diagnosis of the perceived problem in his contribution to John-Stewart Gordon's *Smart Technologies and Fundamental Rights*:

> As Simon Kirchin explains, "the key relativistic thought is that the something that acts as a standard will be different for different people, and that all such standards are equally authoritative" (Kirchin 2012, 15). Particularly problematic is the extreme version, which denies there being any moral judgments or standards that could be objectively true or false (in contrast to moderate versions that do admit of a certain degree of objectivity) (Moser and Carson 2001, 3). Given the apparent rejection of any such standard by Coeckelbergh and Gunkel, they seem to be hard-pressed to explain how the radically relational ethics (to use Coeckelbergh's own term (Coeckelbergh 2010b, 218)) that they are advocating avoids the extreme version (Mosakas 2021a, 95).

The perceived problem with relativism (especially the extreme version of it that concerns Mosakas) is that it encourages and supports a situation in which, it seems, anything goes and all things are permitted. But as both Coeckelbergh and I have argued in other contexts (Gunkel 2018; Coeckelbergh 2020), this particular understanding of *relative* is limited and the product of a culturally specific understanding of and expectation for ethics.

Robert Scott (1976), for instance, understands *relativism* entirely otherwise—as a positive rather than negative term: "Relativism, supposedly, means a standardless society, or at least a maze of differing standards, and thus a cacophony of disparate, and likely selfish, interests. Rather than a standardless society, which is the same as saying no society at all, relativism indicates circumstances in which standards have to be established cooperatively and renewed repeatedly" (Scott 1976, 264). Charles Ess (2009, 21) calls this alternative *ethical pluralism*, which he distinguishes from relativism, strictly speaking: "Pluralism stands as a third possibility—one that is something of a middle ground between absolutism and relativism . . . Ethical pluralism requires us to think in a 'both/and' sort of way, as it conjoins both shared norms and their diverse interpretations and applications in different cultures, times, and places." Likewise, Luciano Floridi (2013, 32) advocates a "pluralism without endorsing relativism," calling this third alternative or middle ground *relationalism*. Unfortunately, these efforts at synthesizing third terms, which endeavor to split the difference between the logical opposition that typically distinguishes moral absolutism from relativism, do not (as we have already seen with the mediating third alternative of slavery) go far enough.

Fortunately, there are others, like Rosi Braidotti, who go beyond these efforts to synthesize mediating third alternatives by calling upon and mobilizing "a form of non-Western perspectivism," which exceeds the limits of Western epistemology and axiology. "Perspectivism," as Eduardo Viveiros de Castro (2015, 24; emphasis in original) explains in his work with Amerindian traditions, "is not relativism, that is the affirmation of the relativity of truth, but relationalism, through which one can affirm *the truth of the relative is the relation.*'" For this reason, Braidotti (2019, 90) finds that perspectivism is not just different from but is "the antidote to relativism": "This methodology," she explains, "respects different viewpoints from equally materially embedded and embodied locations that express the degree of power and quality of experience of different subjects." Braidotti therefore recognizes that what is called "truth" is always formulated and operationalized from a particular subject position, which is dynamic, different, and diverse.

The task, then, is not to escape from these differences in order to occupy some fantastic transcendental vantage point but to learn how to take responsibility for these inescapable alterations in embodied perspectives and their diverse social, moral, and material consequences. And in this matter, positive law provides a useful template and prototype insofar as legal formulations are social constructs relative to specific times and places. Unlike heady metaphysical speculation about who is a person and what is a thing, law needs to make decisions—literally, a cut in the fabric of being—to divide between who counts as a legitimate legal subject and what does not as an object. The trick is not to find a single, common truth that is universally valid in all times and places, but to recognize how any proposed "universal truth" is itself already part of the same game—a particular way of understanding Things that is raised to the standpoint of a universal through the imposition of power. The ethics of Things, therefore, does not endorse relativism (as it is typically defined) but embodies and operational-izes perspectivism, which deconstructs the simple binary logic that defines relativism in opposition to moral absolutism. It attends to the specific exercises of power and the politics of ethics instead of covering these over with the façade of naturalness and universality.

2. Performative contradiction. The other common objection, one that was initially formulated and mobilized by Jürgen Habermas (1987, 185), is that deconstruction engages in *performative contradiction*—that is, situations where word (what is said) and deed (how it is asserted) seem to be at odds with each other. In the domain of robot rights and relational ethics, this objection is something that has been introduced and developed by Henrik Skaug Sætra in an essay titled "Challenging the Neo-Anthropocentric Relational Approach to Robot Rights." "Relationalism," Sætra (2021, 1) explains, "purportedly opens the door for considering robot rights and moving past anthropocentrism. However, I argue that relationalism is, quite to the contrary, a form of neo-anthropocentrism that re-centers human beings and their unique ontological properties, perceptions, and values."

The critical target in this effort is anthropocentrism or (better stated) its opposite. According to Sætra's reading of the literature, relationalism promotes itself as being nonanthropocentric but actually is anthropocentric in practice. As Sætra (2021, 6) explains: "My first objection is that relationalism is arguably deeply anthropocentric because moral standing is derived exclusively from how human beings perceive and form relations with other entities. As we have seen, moral standing is here derived from how something is treated, and not what it is. This means that humans are key to determining value, as it is how entities are treated and perceived by humans that determine their moral standing."

There are two responses to this objection. First, humans are key insofar as ethics and law are human endeavors, or what María Puig de la Bellacasa (2017, 2) calls *human trouble*. But that does not mean that they are exclusively concerned with or limited to human-only matters. The relational turn, like all forms of what Donna Haraway (1991) calls *situated knowledge*, comes from somewhere and is embedded and embodied in specific subject positions. To expect that any form of knowing would be able to escape from these human (all too human) conditions of possibility and operate from some superhuman position of transcendental objectivity is a metaphysical fantasy that is reserved for the gods. In other words, the axiological purity that Sætra operationalizes as a kind of litmus test is a metaphysical fiction. So yes, the relational turn or relational ethics, like all moral theories and ethico-political practices, is beholden to human concerns, perceptions, and values. And like all sciences or modes of knowing, the critical task is not and cannot be to escape from these existential conditions but to learn how to respond to and to take responsibility for them.

Second, Sætra is exactly right: relationalism is *not* nonanthropocentric. But he is incorrect in concluding that this double negative implies a positive—namely, that it is anthropocentric. Nonanthropocentric ethical theories, as Sætra characterizes and explains, include a number of moral innovations that aim to decenter human exceptionalism and culminate, for him at least, in ethical biocentrism: "As compared to the previous type of non-anthropocentrism, ethical biocentrism does not require us to uncover, or conjure up, the interests, preferences, etc., of other entities. Instead, they are considered valuable just because of being what they are, which is why the terms intrinsic or inherent value are often used" (Sætra 2021, 5). But this is not really all that different. Like the anthropocentric model that it contests, ethical biocentrism is still an ontology-driven transaction, where what something is—its being what it is—determines its intrinsic or inherent value. The problem, then, is not just with anthropocentrism or its negation, but with any and all "epistemic centrisms"—which, as Janina Loh (2021, 109) points out, "remain committed to the paternalism implicit in the subject-object dichotomy."

The relational alternative does not play by these rules. It deliberately flips the script on this entire metaphysical transaction. Following the innovations of Levinas (1969, 304), who famously overturned two-thousand-plus years of Western philosophy by proclaiming that ethics is *the* first philosophy, the relational turn puts the social relationship *first* in terms of both sequence and status. Or as Barad (2007, 139) describes it, the primary unit "is not independent objects with inherent boundaries and properties," but relations—"relations without preexisting relata." This fundamental change in perspective produces something outside the orbit of either anthropocentrism or its nonanthropocentric others (e.g., animocentrism, biocentrism, ontocentrism), producing an "eccentric moral theory" (Gunkel 2018, 164) that deconstructs the very difference that distinguishes and differentiates anthropocentrism from its various alternatives.

This leads to the second instance or occurrence of performative contra-diction, which concerns the status and function of properties. "My second objection," Sætra (2021, 7) writes, "is that relationalism is in reality a cam-ouflaged variety of the properties-based approach. This is so because how we relate to other entities is determined by the properties of these others." In other words, the relational turn can say that it puts relations before relata and makes determinations about moral status dependent on "how some-thing is treated, and not what it is" (Sætra 2021, 6). But this is just patently false because properties still matter. "How we relate to someone, and how an entity acts, is dependent on their properties." Again, Sætra is right, but not for the right reasons.

Properties do play a role in moral decision-making, and they can be a useful and expedient heuristic for responding and taking responsibility in the face of others. What is at issue, then, is not their importance but their function. As Gellers points out, properties are not antithetical to or excluded from relationalism; they are just recontextualized and understood in relational terms. "Coeckelbergh," Gellers (2020, 19) explains, "does not foreclose the possibility that properties may play a role in a relational approach to moral consideration. Instead, he leaves room for 'properties-as-they-appear-to-us within a social-relational, social-ecological context' (Coeckelbergh 2010, 219)." In other words, the properties that are deter-mined to belong to an entity are actually a phenomenal effect of the rela-tionship and not an antecedent ontological condition and cause.

This inverts the usual order of things. In moral philosophy—at least its standard Western varieties—what something is commonly determines how it ought to be treated. According to this largely unchallenged stan-dard operating procedure, the question concerning the status of others—whether they are someone who matters or something that does not—is entirely dependent on and derived from what they are and what capabili-ties they possess (or do not possess). Ontology, therefore, is *first* in both pro-cedural sequence and status. Sætra not only endorses this way of thinking but normalizes and naturalizes it, even though it is the product of a specific philosophical tradition and culture.

The relational alternatives (which should be written in the plural to indi-cate that there is not one alternative but a multiplicity of different versions of this alternative) not only challenge this way of thinking but deliber-ately reverse its procedure. This does not diminish the role of properties;

it simply inverts the direction of the derivation. The morally significant properties—those ontological criteria that had been assumed to ground decisions regarding moral respect—are actually what Slavoj Žižek (2008, 209) calls "retroactively (presup)posited" as the result of and as justification for prior decisions made in the face of social involvements and interactions with others. Consequently, even before we know anything at all about what something is in its essence, we have already been called upon and obligated to make a decisive response.[7]

To give it a Kantian spin, we can say that what something is in itself—*das Ding an sich*—is forever inaccessible insofar as all we ever have access to is how something appears to be relative to us. Whatever we think it is in itself is the result of something we project onto or into it after the fact. So it is not accurate to conclude that relationalism is in reality a camouflaged variety of the properties-based approach. Such a conclusion is possible if and only if one normalizes and naturalizes the standard derivation of *ought* from *is*. It is just as likely—and maybe even more epistemologically honest—to conclude that what is actually an effect of embedded and embodied interactions with others has been mistakenly dressed up as and is masquerading as a cause.

7.2.2 Outcomes and Future Opportunities

In the end, all of this obviously generates more questions than it provides definitive answers. And this might (at least initially) be experienced as something disappointing or even aggravating. You have perhaps come to this book wanting answers to what seemed to be important and timely questions: Are robots just things? Can they be persons? Or should they be characterized and understood as something like a slave, occupying an in-between position that is both/and? And rather than answering these questions with a simple "yes" or "no" and taking sides in the existing disputes, the book has identified, documented, and questioned the standard operating presumptions and shared values of the debate itself. So instead of answering the questions and providing the analysis with some sense of finality or closure, we appear to be caught in a kind of proliferation of inquiry, or what Derrida (1981, 42) had called an *interminable analysis*. Although this is something that clearly cuts against the grain of common sense and frustrates expectations for what one might think a book should do, it is necessary if the deconstruction of Things is to be at all successful, consistent, and rigorously applied. This way of concluding—this arguably inconclusive conclusion—has a number of important consequences and opportunities for future research.

First, although this outcome cuts across the grain of the usual set of expectations, it is a necessary and unavoidable aspect of the philosophical enterprise. The prototypical philosopher, Socrates, does not get himself in trouble for proclaiming inconvenient truths or peddling fake news. He gets himself in trouble with his fellow citizens and is eventually put to death for asking questions and doing so to the point that the inquiry becomes both annoying and uncomfortable (Plato 1982, 23a). Like an insatiable toddler, Socrates heaps question upon question upon question.

Since Socrates, philosophers of different stripes, affiliations, and backgrounds have characterized the task of philosophy in similar ways. "I am," Daniel Dennett (1996, vii) explains, "a philosopher, not a scientist, and we philosophers are better at questions than answers. I haven't begun by insulting myself and my discipline, in spite of first appearances. Finding better questions to ask, and breaking old habits and traditions of asking, is a very difficult part of the grand human project of understanding ourselves and our world." Slavoj Žižek (2006, 137) provides something similar: "There are not only true or false solutions, there are also false questions. The task of philosophy is not to provide answers or solutions, but to submit to critical analysis the questions themselves, to make us see how the very way we perceive a problem is an obstacle to its solution."

Consistent with this effort, deconstruction does not seek to provide definitive answers or solutions to existing problems. It seeks to demonstrate how the way we conceive of and talk about a problem is already a problem and a potential obstacle to developing a solution. Asking whether robots, AI systems, and other Things are things, persons, or something else, like slaves, seems like the right place to begin. But this line of inquiry already determines what can be asked about, who it can be asked of, and what kinds of answers will count and register as appropriate. In addition, this entire ontological order is culturally specific and patronizes a particular way of thinking about and ordering Things. Challenging these questions and their mode of inquiry not only provides the opportunity to identify and address what has been marginalized and excluded from the usual way of proceeding but also opens onto previously unexamined alternatives for thinking about and arranging Things differently, and doing so in a way that can make a real and substantive difference for both persons and things.

Second, there are crucial ethical consequences and political concerns associated with this effort. "The growing self-reflexivity of theory," as Mark Taylor (1997, 325) explains, "seems to entail an aestheticizing of politics that makes cultural analysis and criticism increasingly irrelevant." This is an important criticism, and it is something that has been directly asserted by critics of the existing research in robot rights, who find these theoretical musings and "thought experiments" to be woefully disengaged from the exigencies of real social and political struggles: "The debate about robot rights diverts moral philosophy away from the pressing matter of the oppressive use of AI technology against vulnerable groups in society" (Birhane and van Dijk 2020). In other words, the problem with an "intellectual exercise," like the question of robot rights, the debate between its Critics and Advocates, or even the deconstruction of the person/thing dichotomy is that these efforts, as they become more and more involved in their own questions and problematics, appear to be increasingly cut off from the real problems of real people and the things that really matter. "Instead of engaging the 'real,'" Taylor (1997, 207) concludes, "theory seems caught in a hall of mirrors from which 'reality' is 'systematically' excluded."

Although this critique has a certain intuitive appeal—one that plays especially well in the context of populist politics and a culture of anti-intellectualism that is increasingly hostile to anything called or associated with *theory*—it actually misses what is important. Binary oppositions, like that which divides all of existence into the mutually exclusive categories of person and thing, are not just a matter of discursive difference and academic logic; they are the site of real social, political, and moral power. Binary oppositions—wherever they occur and however they come to be arranged—have very real and potentially devastating consequences. As Donna Haraway (1991, 177) explains, "certain dualisms have been persistent in Western traditions; they have been systemic to the logics and practices of domination of women, people of color, nature, workers, animals—in short, domination of all constituted as others, whose task it is to mirror the self."

Conceptual opposites, like that which distinguishes persons from things, do not institute an equitable division between two terms that are on equal footing and of comparable status. They are always and already hierarchical arrangements that are structurally biased. And it is this skewed hierarchical order, as many feminists, environmentalists, postcolonial scholars, and queer theorists have demonstrated and documented, that installs, underwrites, and justifies systems of inequality, domination, and prejudice. As Jinthana Haritaworn writes in their contribution to "Theorizing Queer Inhumanisms": "It is thus essential to interrogate the nonhuman alongside the dehumanization of 'Man's human Others' and to understand what disposes them to becoming animal's other (or object's other)" (Muñoz et al. 2015, 212). There are, then, real moral and political reasons to question the existing order of things and to attempt to operate in excess of and beyond the usual and inherited arrangements. As Hannah Arendt (2018, 461) wrote, "We all grow up and inherit a certain vocabulary. We then have got to examine this vocabulary." And that critical self-reflection is as much a political action as it is a matter of moral and epistemological speculation.

Finally, and most importantly, the deconstruction of things provides a way to intervene in the person/thing dichotomy and thereby move the conversation about robots and other Things beyond the seemingly irresolvable debate and impasse in which we currently find ourselves. It offers another way to respond to and to take responsibility for Things that are neither persons nor things, challenging the hegemony of this way of thinking about and organizing all that is. It therefore opens onto other kinds of ontological orderings that are oriented otherwise and that can accommodate others—not only other human beings, but nonhuman animals, the natural Things of the environment, and a myriad of artificial Things, like robots and AI systems. It is, then, with Things—especially with these technological Things—that we have a chance to think differently and otherwise about ourselves and others.

This means that the question concerning robot rights or the moral and legal status of AI systems is not really—or not exclusively—about the artifacts. It is about us and the limits of who is included in and what comes to be excluded from that first-person-plural pronoun *we*. It is about how we decide—together and across differences—to respond to and take responsibility for our shared social reality with others and other kinds of otherness. It is, then, in responding to the opportunities and challenges posed by seemingly intelligent and social artifacts that we are called to take responsibility for ourselves, for our world, and for the other Things—whether naturally occurring or artificially made—that are encountered here. Thus, the question of robot rights or the moral and legal status of AI systems is indeed a *speculative* matter. Not because the debate is focused on future possibilities that might be or could happen, but because it holds up a mirror (*speculum*) to our human, all too human ways of organizing and making sense of Things, challenging us to reflect on the limitations of our conceptual apparatus, its consequences, and our own privileged position in the order of things.

7.3   The Order of Things

In the order of Things, human beings occupy a unique and privileged position. It is we who are bestowed with or have taken it upon ourselves to order and organize all Things, separating them into the categories of persons or things—categories that we have initially instituted and defined. This power and privilege is the exclusive purview of that Thing which we ourselves have designated *Homo sapiens*. But as Spider-Man, that fictional hybrid who is both more and less than human, reminds us, "With great power comes great responsibility."[8]

If we continue to use this privileged position to defend and further cement our hegemony over all Things, parsing all that is into an exclusive us-and-them dichotomy—that is, other persons who count and are worthy of respect versus mere objectified things that are there solely for our use and enjoyment—then we benefit from this privilege without taking responsibility for it. We have asserted our dominance over all Things—human and nonhuman, living and nonliving, naturally developed and artificially fabricated. We have made sense of the world and organized Things from a position of largely uninterrogated privilege, where we have always already had the power to decide who counts and what does not. We have named, designated, and categorized Things according to our interests, needs, and desires. And in this effort, we have always already decided who is to be included in the first-person-plural pronoun *we* and who or what is not. This way of thinking has worked, and it continues to work to our advantage and self-satisfaction.

What we now see taking shape in the face or the faceplate of the robot, AI system, and other artifact are the systemic restrictions, structural limitations, and power relations of this way of thinking. Our moral and legal ontology appears to be if not broken then at least straining to the point of breaking. There is, as Esposito (2015, 3) describes it, a crack in the dichotomous model that has been used to sort and organize all beings into one of two mutually exclusive categories: person or thing. And *robot* is the name (or perhaps more accurately characterized, one of the names) for this fracture, designating an anomaly or glitch that cannot be easily domesticated by and accommodated to one category or the other.

But it would be impetuous to conclude from this fact that the robot is the cause of this disruption. Instead, it is an artifact and effect of a necessary and unavoidable systemic disturbance that has been at work deconstructing this binary opposition since the time it was initially introduced in the *Institutes* of Gaius. When Čapek coined the term *robot*, he provided a name for a constitutive exception or that "part that has no part" (Rancière, Pangia, and Bowlby 2001), which was already at work overturning and displacing the fundamental terms that had organized and ordered all Things. As Esposito (2015, 137–138) explains in the final chapter of *Persons and Things*: "For an untold time that has yet to end, we have attributed the same superabundant quality to persons that we have taken away from things. The time has come to rebalance relations. But even before doing that, we need to break through the barrier that has divided the world between opposing species. Without denying the disquieting nature of the revolution we are currently undergoing—especially when technology penetrates our bodies, upsetting orders that have existed for thousands of years—the importance of the shift remains. Perhaps for the first time since the disappearance of ancient societies, things have come back to interpellate us directly."

Consequently, the task before us—a task that begins to stand out in relief in the face of Things that challenge us and our "misunderstood anthropocentrism" (Esposito's term)—is to learn to question our privilege and develop critical perspectives on our own sense of exceptionalism. We—we human beings who have been bestowed with or granted to ourselves the power to decide all Things—need to learn how to take responsibility for our privilege, for the damage and harms that it has perpetrated across the centuries, and for the myriad of exclusions and marginalizations that it has instituted and justified.

Questioning privilege from the position of privilege is never an easy undertaking, precisely because it is often rendered transparent by the assumption that it is just normal, somehow part of the natural order of things, and true for all times and places. But these ways of thinking and acting come from somewhere, and they are supported by a particular ideological formation, one that has effectively concealed itself from view by adopting the guise of universality. Yet even here (with this critical self-reflection), Things are not equal, as this privilege is something that has not been evenly distributed. Fortunately, it is in the face of the robot and other Things—Things that resist, for one reason or another, both reification and personification—that we are able to catch a glimpse of our own unique position and status such that we are invited (or compelled) to investigate and take responsibility for its lineage, its current formations, and its possible futures. It is only by challenging ourselves to engage in this critical self-reflection that we can begin to assemble a moral and legal ontology that can respond to and take responsibility in the face of Things.

Ultimately, then, this is not really about robots, AI systems, and other artifacts. It is about us. It is about the moral and legal institutions that we have fabricated to make sense of Things. And it is with the robot—who plays the role of or occupies the place of a kind of spokesperson for Things—that we are now called to take responsibility for this privileged situation and circumstance. What is needed in response is not some forceful reassertion of the usual ways of thinking but a significantly reformulated moral and legal ontology that can scale to the opportunities and the challenges of the twenty-first century and beyond. What this new framework will look like and how it will be developed is the task for thinking and acting from this point forward. Confronting this will be as terrifying and exhilarating as any of the robot uprisings that have been imagined in science fiction. Because getting this right will require nothing less than a thorough rethinking of everything we thought was right, natural, and beyond question.

Notes

Chapter 1

1. As Alan Pottage (2004, 3) explains, "The distinction between persons and things has always been central to legal institutions and procedures. The institutions of Roman law, to the extent that Rome can be taken as the origin of the Western legal tradition, attached persons (*personae*) to things (*res*) by means of a set of legal forms and transactions (*actiones*) which prescribed all of their permissible combinations."

2. According to research conducted by Cindy M. Grimm and Kristen Thomasen (2021), twelve US states have recently (as of 2017) passed legislation for what they call "sidewalk robots." "Several states," they explain, "have granted pedestrian's rights to sidewalk delivery robots, either explicitly or implicitly. In Pennsylvania and Ohio, pedestrian rights are granted implicitly by expanding the definition of Pedestrian to encompass sidewalk robots. In Missouri's proposed Bill and Arizona's law, sidewalk delivery robots are explicitly given the 'rights and duties applicable to a pedestrian under the same circumstances'" (11–12).

3. The two sides of the debate were initially identified in a prescient essay written by Marshall S. Willick and published in *AI Magazine*: "These developments have led to a significant philosophical disagreement which underlies the imminent legal controversy over the classification of AI-equipped computers. Some commentators feel that in defining the scope of legal personality, the danger of an incorrect decision is too great to be stingy. . . . Others, beginning with a predisposition that 'artificial intelligence' is a contradiction in terms, would reject *ab initio* any attempt to recognize computers as persons" (Willick 1983, 13).

4. Deciding on names for the two sides in the debate is not a nominal issue. My initial inclination was to repurpose the terms of political debate, calling one side the "robot right" and the other the "robot left." This determination had two things going for it. First, it highlighted the political nature of the contest—that is, the fact that one side has been more conservative or restrictive in its arguments, while the other side has been more liberal and accommodating of others. Second, it designated

not a simple either/or binary distinction but a spectrum of differences bounded by two extremes that are polar opposites of each other. But, as reviewers of my initial draft correctly pointed out, this choice of terminology, despite any advantages it might have, also had an unfortunate potential to polarize and prejudice readers in advance. Other formulations, coming from the domain of either legal proceedings or formal debate, were also considered. But these were rejected because they are organized in terms of an exclusive binary difference in which one term in the pair is situated as the negative counterpart of the other—for example, plaintiff/defendant, pro/con, or affirmative/negative. *Critic* and *Advocate* were the compromise, and like all compromises, they are functional designations but not necessarily perfect.

5. The terms *robot* and *AI* are often mixed up and substituted for one another within the published literature. A good example of this can be found in Nils J. Nilsson's *The Quest for Artificial Intelligence: A History of Ideas and Achievements* (2010). In this landmark book—"landmark" insofar as it is often considered to be one of the best, if not *the* best, of the comprehensive histories of the science and technology of AI— the two terms are used interchangeably, or at least in a way that does not draw a clear and explicit distinction between the one and the other. Although the book is advertised as a history of AI (the phrase *artificial intelligence* is in the title), Nilsson begins the story with ancient automatons or robots—the Tripods of Hephaistos, Pygmalion's "living" statue Galatea, the numerous robots planned or constructed by Leonardo da Vinci and Jacques de Vaucanson, and the science fiction robots of Čapek's *R.U.R.* and Isaac Asimov's robot stories.

6. For a more complete account and detailed characterization of the method of deconstruction (which is, if you want to be technical about it, not really a "method"), see *Deconstruction* (Gunkel 2021).

7. Immanuel Kant wrote three critiques, the *Critique of Pure Reason* (first published in 1781 and revised in 1787), the *Critique of Practical Reason* (initially published in 1788), and the *Critique of Judgment* (originally published in 1790). For this reason, the critiques are often referred to by number: first critique, second critique, and third critique.

8. *Western*, as opposed to *Eastern* (which is typically conceptualized *via negativa* as "non-Western" and Other), is one more of those seemingly perennial but ultimately troubled and troubling binary oppositions that needs to be submitted to deconstruction. As Nancy S. Jecker (2021, 1) explains: "I refer to various positions as 'Eastern' to indicate a diverse range of views that originated in East, South, and Southeast Asia and are recurrently espoused among people there. I do not mean to suggest that the views in question are held uniformly by people from these regions or to deny that people outside these regions espouse them. Likewise, I refer to various positions as 'Western' to indicate diverse views that originated with the Greeks and are recurrently held by people in the Americas, most European countries, and Australasia. I do not mean to imply that all people living in these regions hold such views or that

no one outside these regions holds them." In other words, Western and Eastern—like any of the other conceptual dichotomies that structure both language and thought—provide handy conceptual distinctions for characterizing different intellectual traditions, even though they are, and we know that they are, technically incorrect and imprecise.

9. The first-person-plural pronoun "we" marks a boundary between who is to be included in that collective subject position and what remains excluded from membership. In deploying the pronoun, the one speaking either makes an implicit assumption about who is addressed by and included in this subject position or issues a call that comes to be recognized by others as speaking to and involving them. The former is an *implication* proceeding from the one who makes the statement; the latter can be understood, following Louis Althusser (2001, 85–126), as *interpellation* whereby recipients come to recognize themselves as having been hailed.

Chapter 2

1. Within the Western philosophical tradition, the point of contact for this is Immanuel Kant. In his influential work *Critique of Pure Reason*, Kant famously distinguished the *object*, which appears to us through the mediation of our senses, from the thing-in-itself, which remains always and forever inaccessible and remote. An object, or what is called *Gegenstand* (literally "standing or placed against") in Kant's German, takes place and has its place opposite a knowing human subject. For this reason, things are not, in and of themselves, objects; they become objects by being situated in opposition to a subject. Or as Heidegger (2012, 37) aphoristically expresses it in the third of the *Bremen Lectures*: "What stands over against is the object for the subject."

2. Verbal constructions like *see, seeing through*, and *transparency* are all formulated in terms of optics and the sense of sight. Such constructions could and should be critically reassessed for their ableist assumptions and consequences. But it also should be noted that these metaphorical formulations are themselves part and parcel of the Western philosophical tradition. As an example, one only needs to recall the role that light, shadows, and vision play in a foundational text like the Allegory of the Cave situated at the center of Plato's *Republic*.

3. *Wizard of Oz* is a term utilized in human-computer interaction (HCI) and human-robot interaction (HRI) studies to describe experimental procedures in which test subjects interact with a computer system or robot that is assumed to be autonomous but is actually controlled by an experimenter who remains hidden from view. The term was initially introduced by John F. Kelly in the early 1980s.

4. If we use the data collected by Harris and Anthis (2021), the number of publications advancing rights for robots ("rights" in this case taken as including both moral and legal rights) exceeds those arguing in favor of the status quo by a factor

of seven. Harris and Anthis categorize and rate published documents using a Likert-esque scale, where 1 indicates that the document "argues forcefully against consideration, e.g. suggesting that artificial beings should never be considered morally," and 5 indicates that the document "argues forcefully for consideration, e.g. suggesting that artificial beings deserve moral consideration now." If we discard median scores of 3 and those that are rated NA, the data provides evidence of 148 publications arguing in support of moral consideration for robots and twenty-one publications arguing for the opposite position—that is, robots never having moral status.

5. This method, as Coeckelbergh effectively demonstrates in *Growing Moral Relations: Critique of Moral Status Ascription* (2012), is rather straightforward and intuitive: identify one or more morally relevant properties and then find out if the entity in question has them or would be capable of having them (or not). Following this line of reasoning, determining whether something, like a robot or AI, could be categorized as a moral or legal subject or whether it is just an object would be a rather simple and straightforward undertaking, proceeding by way of three steps:

1. Having property P is sufficient for moral status S.
2. Entity E has property P.
3. Entity E has moral status S.

In other words, *we* (and who is included in this first-person plural pronoun is not without consequences) first make a determination as to what ontological property or set of properties we believe are sufficient for something to have independent moral and/or legal status that would need to be taken into account and respected. In effect, we identify what are the qualifying criteria that would be needed for "something" to be recognized as "someone" and not just an instrument or thing. We then investigate whether a particular entity, that is, a robot or an AI, actually possess that property or set of properties (or not). Finally, and by applying the criteria decided in step one to the entity identified in step two, we can "objectively" determine whether the artifact in question either can or cannot have a claim to moral status and/or the protections of rights.

6. In February 2022, Ilya Sutskever, who at the time was the chief scientist at OpenAI, posted the following to the Twitter social media platform: "it may be that today's large neural networks are slightly conscious" (see https://twitter.com/ilyasut /status/1491554478243258368). The statement triggered a furor of activity, with respondents not only questioning what the phrase *slightly conscious* could possibly mean but also vehemently arguing against the claim that large neural networks were capable of achieving anything approaching consciousness

7. Modified versions of a Turing test used for addressing questions of machine moral agency and patiency have been proposed by Robert Sparrow, who discussed a "Turing Triage Test" (2004), and by Colin Allen, Gary Varner, and Jason Zinser (2000) and Anne Gerdes and Peter Øhrstrøm (2015), who discussed a "moral Turing test."

Chapter 3

1. According to Anke Graness (2018), there are, within the Western tradition, two ways that the individual comes to possess or to be in possession of the powers that make one a person:

   1. Ontological personalism, whereby personhood is understood to be "an immortal essence of all human beings." "In this view, personhood is associated with an inviolable dignity that merits unconditional respect. This stance is predominantly found in Christian theology" (40). And its organizing principal is the *Imago Dei*.

   2. Moral education, whereby the qualities that make one a person are "not given merely by membership in the species *homo sapiens* . . . [but] acquired by moral and cultural education" (40). This view is more in line with developments in modern European, Enlightenment philosophy, like that developed by the German philosopher Immanuel Kant, who Graness identifies as one of the most, if not *the* most, famous representatives of this way of thinking. For Kant, being a moral person "is not the property of a human being qua member of the human species, but a trait linked to specific characteristics and abilities" (41) that need to be cultivated through *Bildung*, a German word that means both "education" and "enculturation."

Irrespective of how it comes to be and is possessed—either by way of an ontological gift or through the process of careful and deliberate cultivation—Western theories of the person are centered in and concern the individual.

2. Ubuntu has, for better or worse, become the go-to alternative in these discussions and debates. This is a positive development to the extent that turning to these sub-Saharan African traditions and texts provides an important challenge to the often unquestioned hegemony of Western, European philosophy and its canon, which, quite honestly, reads like a who's who of dead white cisgender males. But it is a potential problem, because the inclusion and appropriation of these other ways of thinking risk reproducing the original sin of European colonialism, something Edward Said called *Orientalism*, whereby the exotic Other becomes, as bell hooks (1992, 21) described it, a kind of spice or "seasoning that can liven up the dull dish that is mainstream white culture." Turning to the resources that are available in non-Western traditions, texts, and theories always runs the risk of falling into this trap. But that exigency is no excuse for not engaging in productive dialogue with others. The critical task is to learn how best to respond to others and to take responsibility for the modes of response. This is what is designated by the term *ethics*.

3. As Tomasz Pietrzykowski (2018, 7) remarks in the context of trying to sort out the legal concept of "person": "Indeed, it may be a feature of all basic concepts, not only legal ones, that their internal complexity is not revealed until some commonly

accepted definitions, explanations and approaches widely adopted to elucidate them are themselves turned into objects of critical reflection and analysis."

4. In Christianity, "the *Imago Dei* remains an important, yet elusive topic in theological anthropology" (De Cruz and Maeseneer 2014, 95), and there have been several competing and influential interpretations of the concept. These different renderings turn on how the Latin word *imago*—derived from the Hebrew צלם and typically described as shape, resemblance, figure, shadow—is translated and operationalized. Despite important differences, all of the interpretations seek to explain how the properties or qualities that make someone a person come to be imparted or communicated from the divine creator to the human creature. For a more detailed investigation of this subject, especially as it relates to the possible extension of the title of *person* to AI and robots (arguably the image of an image), see Joshua K. Smith's *Robotic Persons: Our Future with Social Robots* (2021a) and *Robot Theology: Old Questions through New Media* (2021b).

5. The conceptual opposition of natural versus nonnatural constitutes another of the fundamental binary dichotomies that need to be submitted to critical reassessment and deconstruction. See, for example, Keekok Lee's *The Natural and the Artefactual* (1999) and Steven Vogel's *Thinking like a Mall* (2015). My thanks to Joshua Gellers for this insight.

6. This is also a good example of how and why the term *legal person* is often preferable to *artificial person*. In cases like this, the river achieves legal recognition of person following the model that had been successfully used with corporations. But unlike a corporation, which is the paradigmatic artificial person, the river is a natural object. Consequently, it is important to distinguish between two different ways in which the word *artificial* applies when used in this context. The corporation is an artificial being and its status as a person is an artifact of law. A river would only be artificial in the second sense. Unlike the corporation, it is a natural object that is recognized as a person due to the artifice of law. Using the term *legal person* has the advantage of not needing to distinguish between these two different senses of the word *artificial*.

Chapter 4

1. The reasons provided have to do with matters of the body and the limitations imposed on these entities because of the mode of their embodiment. The chimpanzee, it is explained, grew up in the weightlessness of space, which has an irreversible effect on the cardiovascular system. For this reason, "she can only live in weightlessness." The same situation is said to be a limiting condition for the AI, which was "assembled of necessity in space and cannot function in earth gravity" (Leiber 1985, 4).

2. The ancient Greek words τέχνη (*tékhnē*) and φύσις (*phúsis*) are often understood and presented as conceptual opposites. The former can be translated as *art*, *craft*, or

skill and is the etymological root of the English word *technology*. The latter refers to that which exists not by art, artistry, or some kind of artificial means but by *nature* (see Dunshirn 2019).

3. Although Goertzel does not reference it directly, this line of argument was initially proposed and prototyped by the American philosopher Hilary Putnam in an article for *The Journal of Philosophy*, published in 1964. In a section titled "Should Robots Have Civil Rights?," Putnam considers the theoretical possibility of constructing robots that are "psychologically isomorphic" to a human being. If and when this is achieved—and Putnam constructs the argument in terms of a conditional statement—then the "civil rights of robots" may become an urgent matter. The idea is taken up and further developed in a few essays published toward the end of the twentieth century. In the first issue of *Social Epistemology*, Chris Fields (1987, 5) offered the following proposal concerning the evolving social status of computers:

> Once computers are intelligent enough . . . they will qualify as members of society, with rights, social responsibilities, and so forth. The motivation behind this view was stated clearly by Hilary Putnam: if a machine satisfies the same psychological theory as a human, then there is no good reason not to regard it as a conscious being in exactly the same way that the human is regarded as a conscious being. If being a conscious being with certain psychological properties is sufficient for inclusion in society, then a machine with those properties deserves inclusion as much as a human with the same properties. This amounts to the position that sufficiently human-like machines would count as artificial persons.

In a text from 1988, Sohail Inayatullah and Phil McNally argue for the "rights of robots" on the condition that future developments in AI technology will eventually produce intelligent artifacts that could be considered rational actors: "Eventually, AI technology may reach a genesis stage which will bring robots to a new level of awareness that can be considered alive, wherein they will be perceived as rational actors. At this stage, we can expect robot creators, human companions and robots themselves to demand some form of recognized rights as well as responsibilities" (128). And in "Legal Personhood for Artificial Intelligence," Lawrence Solum (1992, 1273) makes a similar conditional argument: "If AIs behaved the right way and if cognitive science confirmed that the underlying processes producing these behaviors were relatively similar to the processes of the human mind, we would have very good reason to treat AIs as persons."

4. At one time, when research in this area was relatively new, it was both possible and practical to provide an exhaustive account of the available publications. And this is precisely what was developed for and presented in both *The Machine Question: Critical Perspectives on AI, Robots and Ethics* (Gunkel 2012) and *Robot Rights* (Gunkel 2018). At this time (November 2022), and after the "exponential increase" in research activity documented by Jamie Harris and Jacy Reese Anthis (2021), this is no longer feasible. Therefore, instead of trying to provide an exhaustive account that includes everything, we will limit the analysis to a representative sample, focusing

attention on those texts that either have been recognized as influential (indicated by numbers of subsequent citations in the literature) or provide unique insights that are useful and informative for the analysis.

5. Although not directly cited in the context of this early essay, this way of proceeding—this deliberate inversion of the standard method for deciding questions or moral status—follows the innovations of the twentieth-century Lithuanian/Jewish thinker Emmanuel Levinas. For more on Levinas's potential contribution to the "relational turn" in AI and robot ethics, see my books *The Machine Question* (2012) and *Robot Rights* (2018), as well as subsequent essays from Mark Coeckelbergh, like "The Moral Standing of Machines: Towards a Relational and Non-Cartesian Moral Hermeneutics" (2014) and "Artificial Intelligence, Responsibility Attribution, and a Relational Justification of Explainability" (2020). For a critical reappraisal of this particular use and application of Levinasian philosophy, see Patrick Gamez's "A Friendly Critique of Levinasian Machine Ethics" (2022).

6. The difference is a matter not only of terminology but also of philosophical traditions and a long-standing decision that divides the discipline. Phenomenology as an approach and method tends to be associated with what is called *continental philosophy*, whereas concepts of behaviorism (like many of the other isms in the field) tend to be situated in the analytic/Anglo-American tradition. The difference is obviously arbitrary, culturally specific, and political (i.e., instituted from and in support of particular positions of institutional power). It should not really matter anymore as the differences have never really been substantive, but, for better or worse, it persists.

7. The prototypical example utilized in these arguments and discussion is almost always a toaster. The selection of this particular kitchen appliance is not accidental nor unimportant. For more on this subject, see Gunkel (2018, 192).

Chapter 5

1. This marks an important shift in the subject. Arguments for (or against) the extension of natural personhood focus on individual entities and their essential capabilities and/or internal states. Argument for (or against) the extension of legal personality focus not on individual subjects (and what goes on inside) but on social context and its functioning. Problems occur when one tries to move between these two domains because they are, quite literally, not talking about the same subject.

2. As Jacob Turner (2019, 174) points out, *legal personality* tends to be used in the context of UK legal proceedings and jurisprudence, whereas *personhood* tends to be the privileged term in the US. We will, following Turner's advice and precedent, use the two terms interchangeably.

3. There is, as nineteenth-century German philosopher G. W. F. Hegel (1977) insightfully pointed out, no presuppositionless mode of scientific inquiry; all analyses

make and proceed from prior assumptions. What makes the difference is whether one recognizes and acknowledges the inherent limitations and challenges of these assumptions.

4. "The arrival of this geological epoch," Gellers (2020, 117) writes, "presents a moment for reflecting on the ways in which modern systems of law and governance have failed to prevent the current environmental crisis. In particular, the Anthropocene calls upon us to question whether an anthropocentric worldview is sustainable, given the havoc it has wrought on the Earth and its inhabitants."

5. It should be added that this effort to tip the scales or influence the outcome is something that can be considered "uncontroversial" only from the perspective of those who would already benefit from the exercise of such prejudice. This is a kind of "fraternal logic," where one stacks the deck in favor of those who already enjoy the privilege of being members of the club.

6. John-Stewart Gordon and Ausrine Pasvenskiene (2021) provide a much-needed critical review of this literature in the essay "Human Rights for Robots? A Literature Review."

7. In advancing this argument, the authors bracket the question of natural or moral personhood. Although they give brief consideration to both the properties approach to deciding questions of moral status and the epistemological complications addressed by the behavioralist litmus tests, they are quick to point out that none of this moral personhood stuff is sufficient for determining legal personality. "Even if robots were to be constructed on the mass scale and to acquire moral rights, this would not fully settle the question of whether the law should recognize them as legal persons. Legal systems are flexible as to what actors they confer legal personality upon, and they need no evidence of supposed inherent qualities of an actor in order to do so" (Bryson, Diamantis, and Grant 2017, 284).

8. It should be noted that there is some equivocation concerning this rather emphatic imperative, which punctuates the final sentence of the essay. At the beginning of the text, Bryson, Diamantis, and Grant (2017, 274) seem to want to hedge their bets by issuing a less determinative position: "In this article, we ask whether a purely synthetic entity could and should be made a legal person. Drawing on the legal and philosophical framework used to evaluate the legal personhood of other non-human entities like corporations, we argue that the case for electronic personhood is weak." According to this initial statement of purpose, then, the article was never intended to actually prove that "purely synthetic intelligent entities should never become persons" but only to demonstrate that the case for extending legal personhood to robots and other artifacts was relatively "weak." But even this formulation is itself a rather weak assertion and ultimately insufficient insofar as the essay fails to provide the proposed cost-benefit analysis that would have proven or supported it. In the end, then, Bryson, Diamantis, and Grant's essay is unable to close

the deal on either the absolute imperative that purely synthetic intelligent entities never become persons or the more measured but relatively weaker claim that the case for electronic personhood is weak.

9. As Chopra and White (2011) explain, their preferred solution to the contracting problem—at least as regards the current state-of-the-art technology—is the *agency doctrine*, whereby the technological artifact is considered an intelligent nonperson actor with limited legal capacity. "The most cogent reason for adopting the agency law approach to artificial agents in the context of the contracting problem is to allow the law to distinguish in a principled way between those contracts entered into by an artificial agent that should bind a principal and those that should not" (44). But this formulation, as they explicitly recognize, has important and potentially disturbing parallels to the legal status of human slaves as established in both Roman and American law. We will take up and examine the opportunities and challenges of this "electronic slave metaphor" (42) in the penultimate chapter.

10. The phrase "answer for . . ." is deliberate in this context. This is because the "concept of responsibility," as the philosopher Paul Ricœur (2007, 12) pointed out is a matter of being *able* to *respond* to or answer for a decision or action. The question in the face of increasingly autonomous technology, then, is this: Who or what can be or should be responsible for the consequences of decisions and actions instituted by robots, AIs, or other autonomous technology? Who or what is able to respond to or answer for what the technological artifact does or does not do? (see Gunkel 2020b).

Chapter 6

1. For a detailed literary analysis of how Shelley's fiction set the stage for contemporary ethical and political debates about artificially intelligent creatures, see Eileen Hunt Botting's *Artificial Life after Frankenstein* (2021).

2. Petersen's proposals have been met with a couple of important and critical responses. In the essay "What's Wrong with Designing People to Serve?," Bartek Chomanski (2019, 993) uses the Aristotelian vice of manipulativeness to argue "that it is unethical to create artificial agents possessing human-level intelligence that are programmed to be human beings' obedient servants." And in "Designing (Artificial) People to Serve—the Other Side of the Coin," Maciej Musiał (2017, 1087) employs Jürgen Habermas's "critique of positive liberal eugenics" to argue that "any kind of intentional designing inevitably wrongs the designed beings regarding their freedom, autonomy, equality and identity."

3. Efforts to address and critique the not-so-hidden racial dimensions of robot slavery, although scarcely identified and dealt with in the legal literature, have been increasingly important and visible in both science fiction studies (Lavender 2011; Ginway 2011; King 2013; Chude-Sokei 2015; Hampton 2015) and the more

philosophically oriented work in robot/AI ethics (Gunkel 2012, 2018, 2020a; Estrada 2020; Gellers 2020).

4. In advancing this concept of limited inclusion, Pietrzykowski's proposal follows and is consistent with the stipulations of animal rights philosophy as developed by both Peter Singer and Tom Regan. According to Regan (1983), for example, the case for animal rights does not include all animals but is limited to those species with sufficient complexity to have at least a minimal level of mental abilities similar to a human being. For this reason, the word *animal* in Regan's *The Case for Animal Rights* is limited to "mentally normal mammals of a year or more" (78) and excludes virtually everything else.

5. It is not insignificant or coincidental that these solutions rely on color imagery. Both Schirmer and Mocanu characterize the standard thing/person dichotomy as a limited black versus white binary and propose in-between solutions that are shades of gray or a spectrum of different colors. The same terminological distinction—black versus white—is also a fundamental component of the concept of slavery as it was instituted in both North and South America, where perceived differences in race divided white masters from Black slaves. The use of color imagery is clearly useful for characterizing the systemic insufficiencies and limitations of binary oppositions. But these metaphors are not neutral, and they come with their own axiological assumptions and social/political consequences. For these reasons, it is difficult to challenge the institutions of slavery by using and relying on logics and rhetorics that are themselves related to and closely associated with systems of hegemony and colonialism. It is this problem—the systemic inequalities that are already hardwired into the very fabric of language itself—that deconstruction seeks to address and remediate.

Chapter 7

1. On the affinities of and points of contact between deconstruction and queer theory, see Nicholas Royle's (2009, 113–134) essay, "Impossible Uncanniness: Deconstruction and Queer Theory" in his book *In Memory of Jacques Derrida*.

2. This formulation is not something that is limited to Heidegger's analysis; it finds similar expression in the work of other twentieth-century thinkers, like the American communication theorist James Carey (1992, 25): "I want to suggest, to play on the Gospel of St. John, that in the beginning was the word; words are not the names for things but, to steal a line from Kenneth Burke, things are signs of words. Reality is not given, not humanly existent, independent of language and toward which language stands as a pale reflection. Rather, reality is brought into existence, is produced by communication—by, in short, the construction, apprehension, and utilization of symbolic forms."

3. This seemingly small but significant typographical alteration—that is, identifying this third, nondialectic term with the name Thing—follows a precedent that

has been developed in the literature of/on deconstruction. In Derrida's own work, deconstruction of the defining logocentric opposition of speech/writing results in a third term that he identifies as *writing*—or, better, *arche-writing*. This term designates another concept of writing that is prior to and outside of the conventional or vulgar concept of writing that has been defined and characterized in opposition to the privilege of speech. Something similar occurs in the work of Levinas, where the word Other—typically written (in translations of Levinas's texts) with the capital O—has been deployed to identify a primordial experience with externality that is anterior to and the prior condition of the subsequent formation of the conceptual pair that opposes *self* to *other*. These nominal repetitions—these repetitions with a difference—are a necessary component of the two-step procedure of deconstruction. But they also (and somewhat regrettably) provide the occasion for misreadings and misunderstandings of the deconstructive effort. For more on this subject, its exigencies, and its consequences, see *Deconstruction* (Gunkel 2021).

4. OOO is not the only development in this direction. There is also new materialism (Coole and Frost 2010), speculative realism (Shaviro 2016), the parliament of things (Latour 2012), and others. For this reason, the brief consideration of OOO included here is simply exemplary and not meant to be an exhaustive account of the wide range of different and interdisciplinary efforts to think about and take care of things. Importantly, what this proliferation of different strategies and approaches indicates is the fact that Western thought is currently engaged in a struggle against the weight of its own traditions, trying to return "to the things themselves" in a way that responds to the phenomenological project that is often attributed to Heidegger's predecessor, Edmund Husserl.

5. In calling upon and referring to the work of thinkers like Heidegger, Levinas, Derrida, Haraway, Benso, Esposito, and so on, it is important to point out that none of these are mobilized as or intended to be a kind of *philosopher ex machina* (a brilliant term provided to me by Mark Coeckelbergh). No one philosopher or theorist provides a ready-made "theory of everything" such that their work can be taken up literally and applied as a kind of religious doctrine. Each contributes something unique that can be added into the mix. In calling upon and using the work of these other thinkers, the goal is not to achieve fidelity to their project as such but to draw upon the resources and insights they—and often times they alone—make available. Elsewhere, I have explained this way of proceeding and related it to the efforts of the remix artist or DJ. The composition of a book like this is always a matter of carefully selecting source material (a.k.a. "crate diving"), isolating and sampling different elements and passages, and then assembling and recombining all of this into a new whole that is greater than the sum of its parts. See *Of Remixology: Ethics and Aesthetics after Remix* (Gunkel 2016).

6. The phrase *the face of the Other*, which does so much of the heavy lifting in Levinas's own texts and publications, is open to a number of different readings and

competing interpretations. For a review and critical engagement with the multifaceted aspects of the concept, see *The Changing Face of Alterity: Communication, Technology, and Other Subjects* (Gunkel, Filho, and Mersch 2016).

7. For this reason, relations are neither an ontological criterion nor an epistemic category. They are the prior ethical condition. This is why, for Levinas (1969, 304) and others who have followed his lead, it is ethics, and not ontology or epistemology, that is designated as "first philosophy."

8. This statement, which has often been directly attributed to the Marvel Comics superhero Spider-Man, was initially voiced by Benjamin Parker (a.k.a. Uncle Ben) in conversation with Peter Parker (a.k.a. Spider-Man). My thanks to both Joshua Gellers and Joshua Smith for this important literary clarification.

References

Adams, Brian, Cynthia L. Breazeal, Rodney Brooks, and Brian Scassellati. 2000. "Humanoid Robots: A New Kind of Tool." *IEEE Intelligent Systems* 15 (4): 25–31. https://doi.org/10.1109/5254.867909.

Allen, Colin, Gary Varner, and Jason Zinser. 2000. "Prolegomena to Any Future Artificial Moral Agent." *Journal of Experimental & Theoretical Artificial Intelligence* 12 (3): 251–261. https://doi.org/10.1080/09528130050111428.

Allen, Tom, and Robin Widdison. 1996. "Can Computers Make Contracts?" *Harvard Journal of Law and Technology* 9 (1): 25–52. http://jolt.law.harvard.edu/articles/pdf/v09/09HarvJLTech025.pdf.

Althusser, Louis. 2001. *Lenin and Philosophy and Other Essays.* Translated by Ben Brewster. New York: Monthly Review Press.

Ames, R. T. 1988. "Rites as Rights: The Confucian Alternative." In *Human Rights and the World's Religions*, edited by L. S. Rouner, 199–216. South Bend, IN: Notre Dame University Press.

Andrade, Francisco, Paulo Novais, Jose Machado, and Jose Neves. 2007. "Contracting Agents: Legal Personality and Representation." *Artificial Intelligence and Law*, no. 15, 357–373. https://doi.org/10.1007/s10506-007-9046-0.

Andreotta, Adam J. 2021. "The Hard Problem of AI Rights." *AI & Society*, no. 36, 19–32. https://doi.org/10.1007/s00146-020-00997-x.

Aquinas, Thomas. 1945. *Summa Theologica.* In *Basic Writings of Saint Thomas Aquinas*, vol. 1, edited and translated by Anton C. Pegis. New York: Random House.

Aquinas, Thomas. 2003. *Commentary on Aristotle's Physics.* Translated by Richard J. Blackwell, Richard J. Spath, and W. Edmund Thirlkey. Notre Dame, IN: Dumb Ox Books.

Arendt, Hannah. 1968. *The Origins of Totalitarianism.* New York: Harcourt Books.

Arendt, Hannah. 2018. *Thinking without a Banister: Essays in Understanding 1953–1975*. New York: Schocken Books.

Aristotle. 1944. *Politics*. Translated by H. Rackham. Cambridge, MA: Harvard University Press.

Arkin, Ronald. 1998. *Behavior Based Robotics*. Cambridge, MA: MIT Press.

Asaro, Peter. 2012. "A Body to Kick, but Still No Soul to Damn: Legal Perspectives on Robotics." In *Robot Ethics: The Ethical and Social Implications of Robotics*, edited by Patrick Lin, Keith Abney, and George A. Bekey, 169–186. Cambridge, MA: MIT Press.

Ashrafian, Hutan. 2015. "Artificial Intelligence and Robot Responsibilities: Innovating beyond Rights." *Science and Engineering Ethics* 21 (2): 317–326. https://doi.org/10.1007/s11948-014-9541-0.

Asimov, Isaac. 1976. *The Bicentennial Man and Other Stories*. New York: Doubleday.

Auer-Welsbach, Christoph. 2018. "Fifteen Minutes with Leading #AI Specialist Joanna Bryson." *Medium*, January 9, 2018. https://medium.com/cityai/fifteen-minutes-with-leading-ai-specialist-joanna-bryson-c944b7c3fd25.

Banks, Jamie. 2021. "From Warranty Voids to Uprising Advocacy: Human Action and the Perceived Moral Patiency of Social Robots." *Frontiers in Robotics and AI*, May 28, 2021. https://doi.org/10.3389/frobt.2021.670503.

Barad, Karen. 2007. *Meeting the Universe Halfway: Quantum Physics and the Entanglement of Matter and Meaning*. Durham, NC: Duke University Press.

Basham, Victoria. 2021. "South Africa Issues World's First Patent Listing AI as Inventor." *Global Legal Post*, July 28, 2021. https://www.globallegalpost.com/news/south-africa-issues-worlds-first-patent-listing-ai-as-inventor-161068982.

Basl, John, and Joseph Bowen. 2020. "AI as a Moral Right-Holder." In *The Oxford Handbook of Ethics of AI*, edited by Markus D. Dubber, Frank Pasquale, and Sunit Das, 289–306. New York: Oxford University Press.

Bayern, Shawn. 2015. "The Implications of Modern Business-Entity Law for the Regulation of Autonomous Systems." *Stanford Technology Law Review*, no. 19, 93–112. https://law.stanford.edu/wp-content/uploads/2017/11/19-1-4-bayern-final_0.pdf.

Bayern, Shawn. 2019. "Are Autonomous Entities Possible?" *Northwestern University Law Review*, no. 114, 23–47. https://scholarlycommons.law.northwestern.edu/nulr_online/271/.

Bayley, Barrington J. 1974. *The Soul of the Robot*. Gillette, NJ: Wayside Press.

Beauchamp, Tom L. 1999. "The Failure of Theories of Personhood." *Kennedy Institute of Ethics Journal* 9 (4): 309–324. https://doi.org/10.1353/ken.1999.0023.

Beckers, Anna, and Gunther Teubner. 2021. *Three Liability Regimes for Artificial Intelligence: Algorithmic Actants, Hybrids, Crowds*. Oxford: Hart Publishing.

Bekey, George A. 2005. *Autonomous Robots: From Biological Inspiration to Implementation and Control*. Cambridge, MA: MIT Press.

Benford, Gregory, and Elisabeth Malartre. 2007. *Beyond Human: Living with Robots and Cyborgs*. New York: Tom Doherty.

Benjamin, Ruha. 2019. *Race After Technology: Abolitionist Tools for the New Jim Code*. Cambridge: Polity Press.

Bennett, Belinda, and Angela Daly. 2020. "Recognising Rights for Robots: Can We? Will We? Should We?" *Law, Innovation and Technology* 12 (1): 60–80. https://doi.org/10.1080/17579961.2020.1727063.

Benso, Silvia. 2000. *The Face of Things: A Different Side of Ethics*. Albany, NY: SUNY Press.

Bensoussan, Alain, and Jérémy Bensoussan. 2015. *Droit des Robots*. Bruxelle: Éditions Larcier.

Bentham, Jeremy. 2005. *An Introduction to the Principles of Morals and Legislation*. New York: Oxford University Press. Originally published 1789.

Berg, Jessica Wilen. 2007. "Of Elephants and Embryos: A Proposed Framework for Legal Personhood." *Hastings Law Journal* 59 (2): 369–406. https://repository.uchastings.edu/hastings_law_journal/vol59/iss2/3.

Bertolini, Andrea. 2013. "Robots as Products: The Case for a Realistic Analysis of Robotic Applications and Liability Rules." *Law Innovation and Technology* 5 (2): 214–247. https://doi.org/10.5235/17579961.5.2.214.

Bertolini, Andrea, and Giuseppe Aiello. 2018. "Robot Companions: A Legal and Ethical Analysis." *Information Society* 34 (3): 130–140. https://doi.org/10.1080/01972243.2018.1444249.

Bess, Michael. 2018. "Eight Kinds of Critters: A Moral Taxonomy for the Twenty-Second Century." *Journal of Medicine and Philosophy* 43 (5): 585–612. https://doi.org/10.1093/jmp/jhy018.

Big Think. 2019. "Will Robots Have Rights in the Future? Peter Singer." Big Think. December 14, 2019. YouTube video, 4:13. https://www.youtube.com/watch?v=y_aK_njciZc.

Binder, Otto O. 1957. "You'll Own 'Slaves' by 1965." *Mechanix Illustrated*, January 1957, 62–65. Greenwich, CT: Modern Mechanix Publishing Co.

Birch, Thomas H. 1993. "Moral Considerability and Universal Consideration." *Environmental Ethics* 15:313–332.

Birch, Thomas H. 1995. "The Incarnation of Wilderness: Wilderness Areas as Prisons." In *Postmodern Environmental Ethics*, edited by Max Oelschlaeger, 137–162. Albany, NY: SUNY Press.

Birhane, Abeba, and Jelle van Dijk. 2020. "Robot Rights? Let's Talk about Human Welfare Instead." In *AIES '20: Proceedings of the AAAI/ACM Conference on AI, Ethics, and Society*, 207–213. New York: ACM. https://doi.org/10.1145/3375627.3375855.

Birhane, Abeba, Jelle van Dijk, and Frank Pasquale. 2021. "Debunking Robot Rights Metaphysically, Ethically, and Legally." Paper presented at We Robot, University of Miami School of Law, Miami, FL, September 23–25. https://werobot2021.com /debunking-robot-rights-metaphysically-ethically-and-legally/.

Blackley, Robert. 1984. "Robot Rights Is a Growth Movement." *New York Times*, November 30, 1984. https://www.nytimes.com/1984/11/30/opinion/l-robot-rights-is -a-growth-movement-117430.html.

Boethius. 1860. *Liber de persona et duabus naturis contra Eutychen et Nestorium, ad Joannem Diaconum Ecclesiae Romanae*: 3. *Patrologia Latina* 64, Paris.

Bogost, Ian. 2012. *Alien Phenomenology, or What It's Like to Be a Thing*. Minneapolis: University of Minnesota Press.

Bowyer, Kyle. 2017. "Robot Rights: At What Point Should an Intelligent Machine Be Considered a 'Person'?" The Conversation, February 6, 2017. https://theconversation .com/robot-rights-at-what-point-should-an-intelligent-machine-be-considered-a-person -72410.

Bradgate, R. 1999. "Beyond the Millennium—the Legal Issues: Sale of Goods Issues and the Millennium Bug." *Journal of Information, Law and Technology* 2. http://www2 .warwick.ac.uk/fac/soc/law/elj/jilt/1999_2/bradgate/.

Bradley, Keith. 1994. *Slavery and Society at Rome*. New York: Cambridge University Press.

Braidotti, Rosi. 2019. *Posthuman Knowledge*. Cambridge: Polity Press.

Breazeal, Cynthia L. 2004. *Designing Sociable Robots*. Cambridge, MA: MIT Press.

Brooks, Rodney A. 2002. *Flesh and Machines: How Robots Will Change Us*. New York: Pantheon Books.

Brożek, Bartosz, and Marek Jakubiec. 2017. "On the Legal Responsibility of Autonomous Machines." *Artificial Intelligence and Law* 25:293–304. https://doi.org/10.1007 /s10506-017-9207-8.

Bryant, Levi R. 2011. *The Democracy of Objects*. Ann Arbor, MI: Open Humanities Press.

Bryson, Joanna. 2010. "Robots Should Be Slaves." In *Close Engagements with Artificial Companions: Key Social, Psychological, Ethical and Design Issues*, edited by Yorick Wilks, 63–74. Amsterdam: John Benjamins.

Bryson, Joanna. 2015. "Clones Should NOT Be Slaves." *Adventures in NI* (blog), October 4, 2015. https://joanna-bryson.blogspot.com/2015/10/clones-should-not-be-slaves.html.

Bryson, Joanna, Mihailis E. Diamantis, and Thomas D. Grant. 2017. "Of, for, and by the People: The Legal Lacuna of Synthetic Persons." *Artificial Intelligence and Law* 25:273–291. https://doi.org/10.1007/s10506-017-9214-9.

Bryson, Joanna, and Alan Winfield. 2017. "Standardizing Ethical Design for Artificial Intelligence and Autonomous Systems." *Computer* 50 (5): 116–119. https://doi.org/10.1109/MC.2017.154.

Callicott, J. Baird. 1989. *In Defense of the Land Ethic: Essays in Environmental Philosophy*. Albany, NY: SUNY Press.

Calo, Ryan. 2015. "Robotics and the Lessons of Cyberlaw." *California Law Review* 103 (3): 513–563. http://www.californialawreview.org/wp-content/uploads/2015/07/Calo_Robots-Cyberlaw.pdf.

Calverley, David. 2008. "Imagining a Non-biological Machine as a Legal Person." *AI & Society* 22 (4): 523–537. https://doi.org/10.1007/s00146-007-0092-7.

Čapek, Karel. 2009. *R.U.R. (Rossum's Universal Robots)*. Translated by D. Wyllie. Gloucestershire: Echo Library. Originally published 1920.

Cappuccio, Massimiliano L., Anco Peeters, and William McDonald. 2020. "Sympathy for Dolores: Moral Consideration for Robots Based on Virtue and Recognition." *Philosophy & Technology*, no. 33, 9–31. https://doi.org/10.1007/s13347-019-0341-y.

Carey, James W. 1992. *Communication as Culture: Essays on Media and Society*. New York: Routledge.

Čerka, Paulius, Jurgita Grigienė, and Gintarė Sirbikytė. 2017. "Is It Possible to Grant Legal Personality to Artificial Intelligence Software Systems?" Computer Law and Security Review 33 (5): 685–699. https://doi.org/10.1016/j.clsr.2017.03.022.

Chalmers, David J. 2010. *The Character of Consciousness*. Oxford: Oxford University Press.

Chesterman, Simon. 2021. *We, the Robots? Regulating Artificial Intelligence and the Limits of Law*. Cambridge: Cambridge University Press.

Chocano, Carina. 2017. "What I Care About Is Important. What You Care About Is a 'Distraction.'" *New York Times Magazine*, September 26, 2017. https://www.nytimes.com/2017/09/26/magazine/what-i-care-about-is-important-what-you-care-about-is-a-distraction.html.

Chomanski, Bartek. 2019. "What's Wrong with Designing People to Serve?" *Ethical Theory and Moral Practice* 22 (4): 993–1015. https://doi.org/10.1007/s10677-019-10029-3.

Chopra, Samir, and Laurence F. White. 2004. "Artificial Agents—Personhood in Law and Philosophy." In *ECAI'04: Proceedings of the 16th European Conference on Artificial Intelligence*, edited by Ramon López de Mántaras and Lorenza Saitta, 635–639. Amsterdam: IOS Press.

Chopra, Samir, and Laurence F. White. 2011. *A Legal Theory for Autonomous Artificial Agents*. Ann Arbor: University of Michigan Press.

Chu, Seo-Young. 2010. *Do Metaphors Dream of Literal Sleep? A Science-Fictional Theory of Representation*. Cambridge, MA: Harvard University Press.

Chude-Sokei, Louis. 2015. *The Sound of Culture: Diaspora and Black Technopoetics*. Middleton, CT: Wesleyan University Press.

Churchland, Paul M. 1999. *Matter and Consciousness*. Cambridge, MA: MIT Press.

Ci, Jiwei. 1999. "The Confucian Relation Concept of the Person and Its Modern Predicament." In *Personhood and Health Care*, edited by David C. Thomasma, David N. Weisstub, and Christian Hervé, 149–164. Dordrecht: Springer. https://doi.org/10.1007/978-94-017-2572-9_14.

Clapham, Andrew. 2007. *Human Rights: A Very Short Introduction*. New York: Oxford University Press.

Clark, David. 2001. "Kant's Aliens: The Anthropology and Its Others." *CR: The New Centennial Review* 1 (2): 201–289. https://www.jstor.org/stable/41949284.

Clynes, Manfred E., and Nathan S. Kline. 1960. "Cyborgs and Space." *Astronautics*, September 1960, 26–27, 74–76. Reprinted in *The Cyborg Handbook*, edited by Chris Hables Gray, 29–34. New York: Routledge, 1995.

Cobb, T. R. R. 1968. *Law of Negro Slavery in the United States of America*. New York: Negro Universities Press.

Coeckelbergh, Mark. 2010a. "Moral Appearances: Emotions, Robots, and Human Morality." *Ethics and Information Technology* 12 (3): 235–241. https://doi.org/10.1007/s10676-010-9221-y.

Coeckelbergh, Mark. 2010b. "Robot Rights? Towards a Social-Relational Justification of Moral Consideration." *Ethics and Information Technology* 12 (3): 209–221. https://doi.org/10.1007/s10676-010-9235-5.

Coeckelbergh, Mark. 2012. *Growing Moral Relations: Critique of Moral Status Ascription*. New York: Palgrave MacMillan.

Coeckelbergh, Mark. 2014. "The Moral Standing of Machines: Towards a Relational and Non-Cartesian Moral Hermeneutics." *Philosophy and Technology*, no. 27, 61–77. https://doi.org/10.1007/s13347-013-0133-8.

Coeckelbergh, Mark. 2020. "Artificial Intelligence, Responsibility Attribution, and a Relational Justification of Explainability." *Science and Engineering Ethics*, no. 26, 2051–2068. https://doi.org/10.1007/s11948-019-00146-8.

Commonwealth of Virginia. 2021. "Personal Delivery Vehicles." Virginia Code § 46.2-908.1:1. https://law.lis.virginia.gov/vacode/title46.2/chapter8/section46.2-908 .1:1/.

Coole, Diana, and Samantha Frost. 2010. *New Materialisms: Ontology, Agency, and Politics*. Durham, NC: Duke University Press.

Danaher, John. 2016. "Robots, Law and the Retribution Gap." *Ethics and Information Technology* 18 (4): 299–309. https://doi.org/10.1007/s10676-016-9403-3.

Danaher, John. 2017. "Should Robots Be Granted the Status of Legal Personhood?" *Philosophical Disquisitions* (blog), October 25, 2017. https://philosophicaldisquisitions .blogspot.com/2017/10/should-robots-be-granted-status-of.html.

Danaher, John. 2020. "Welcoming Robots into the Moral Circle: A Defence of Ethical Behaviourism." *Science and Engineering Ethics*, no. 26, 2023–2049. https://doi.org /10.1007/s11948-019-00119-x.

Darling, Kate. 2012. "Extending Legal Protection to Social Robots." *IEEE Spectrum*, September 10, 2012. http://spectrum.ieee.org/automaton/robotics/artificial -intelligence/extending-legal-protection-to-social-robots.

Darling, Kate. 2016. "Extending Legal Protection to Social Robots: The Effects of Anthropomorphism, Empathy, and Violent Behavior toward Robotic Objects." In *Robot Law*, edited by Ryan Calo, A. Michael Froomkin, and Ian Kerr, 213–231. Northampton, MA: Edward Elgar.

Darling, Kate. 2021. *The New Breed: What Our History with Animals Reveals about Our Future with Robots*. New York: Henry Holt and Company.

Datteri, Edoardo. 2013. "Predicting the Long-Term Effects of Human-Robot Interaction: A Reflection on Responsibility in Medical Robotics." *Science and Engineering Ethics* 19 (1): 139–160. https://doi.org/10.1007/s11948-011-9301-3.

Dawes, James. 2020. "Speculative Human Rights: Artificial Intelligence and the Future of the Human." *Human Rights Quarterly* 42: 573–593. https://doi.org/10.1353/hrq .2020.0033.

De Cruz, Helen, and Yves Maeseneer. 2014. "The *Imago Dei*: Evolutionary and Theological Perspectives." *Zygon* 49 (1): 95–100. https://doi.org/10.1111/zygon.12064.

De Pagter, Jesse. 2021. "Speculating about Robot Moral Standing: On the Constitution of Social Robots as Objects of Governance." *Frontiers in Robotics and AI*, December 2, 2021: 1–12. https://doi.org/10.3389/frobt.2021.769349.

Delvaux, Mady. 2016. *Draft Report, with Recommendations to the Commission on Civil Law Rules on Robotics, 2015/2103(INL)*. Committee on Legal Affairs. European Parliament. https://www.europarl.europa.eu/doceo/document/JURI-PR-582443_EN.pdf?redirect.

Dennett, Daniel C. 1996. *Kinds of Minds*. New York: Basic Books.

Dennett, Daniel C. 1998. "Conditions of Personhood." In *Brainstorms: Philosophical Essays on Mind and Psychology*, 267–285. Cambridge, MA: MIT Press.

Derrida, Jacques. 1981. *Positions*. Translated by Alan Bass. Chicago: University of Chicago Press.

Derrida, Jacques. 1993. *Limited Inc*. Translated by Samuel Weber and Jeffrey Mehlman. Evanston, IL: Northwestern University Press.

Derrida, Jacques. 2005. *Paper Machine*. Translated by Rachel Bowlby. Stanford, CA: Stanford University Press.

Díez Spelz, Juan Francisco. 2021. "¿Robots con Derechos? La Frontera Entre lo Humano y lo No-Humano. Reflexiones Desde la Teoría de los Derechos Humanos." *Revista IUS* 15 (48): 1–34. https://doi.org/10.35487/rius.v15i48.2021.742.

Douglass, Frederick. 2018. *Narrative of the Life of Frederick Douglass, an American Slave*. Petersborough, Ontario: Broadview Press. Originally published 1845.

Dunshirn, Alfred. 2019. "Physis." In *Online Encyclopedia Philosophy of Nature/Online Lexikon Naturphilosophie*, edited by Thomas Kirchhoff. https://doi.org/10.11588/oepn.2019.0.66404.

Dvorsky, George. 2017. "When Will Robots Deserve Human Rights?" *Gizmodo*, June 2, 2017. https://gizmodo.com/when-will-robots-deserve-human-rights-1794599063.

Dyschkant, Alexis. 2015. "Legal Personhood: How We Are Getting It Wrong." *University of Illinois Law Review*, August 31, 2015, 2075–2110. https://www.illinoislawreview.org/print/volume-2015-issue-5/legal-personhood-how-we-are-getting-it-wrong/.

Esposito, Roberto. 2015. *Persons and Things*. Translated by Zakiya Hanafi. Cambridge: Polity.

Ess, Charles. 1996. "The Political Computer: Democracy, CMC, and Habermas." In *Philosophical Perspectives on Computer-Mediated Communication*, edited by Charles Ess, 196–230. Albany, NY: SUNY Press.

Ess, Charles. 2009. *Digital Media Ethics*. Cambridge: Polity Press.

Estrada, Daniel. 2018. "Sophia and Her Critics." *Medium*, June 17, 2018. https://medium.com/@eripsa/sophia-and-her-critics-5bd22d859b9c.

Estrada, Daniel. 2020. "Human Supremacy as Posthuman Risk." *Journal of Sociotechnical Critique* 1 (1): 1–40. https://doi.org/10.25779/j5ps-dy87.

Fan, Ruiping. 2010. *Reconstructionist Confucianism: Rethinking Morality after the West.* New York: Springer.

Feenberg, Andrew. 1991. *Critical Theory of Technology.* New York: Oxford University Press.

Felliu, Silvia. 2001. "Intelligent Agents and Consumer Protection." *International Journal of Law and Information Technology* 9 (3): 235–248. https://doi.org/10.1093/ijlit/9.3.235.

Fields, Chris. 1987. "Human-Computer Interaction: A Critical Synthesis." *Social Epistemology: A Journal of Knowledge, Culture and Policy* 1 (1): 5–25. https://doi.org/10.1080/02691728708578410.

Fiennes, Sophie, dir. 2009. *The Pervert's Guide to Cinema.* Mischief Films/Amoeba Film.

Floridi, Luciano. 2013. *The Ethics of Information.* Oxford: Oxford University Press.

Floridi, Luciano. 2017. "Robots, Jobs, Taxes, and Responsibilities." *Philosophy & Technology* 30 (1): 1–4. https://doi.org/10.1007/s13347-017-0257-3.

Friedman, Cndy. 2020. "Human-Robot Moral Relations: Human Interactants as Moral Patients of Their Own Agential Moral Actions towards Robots." In *Artificial Intelligence Research*, edited by A. Gerber. SACAIR 2021. Communications in Computer and Information Science, vol. 1342. Springer, Cham. https://doi.org/10.1007/978-3-030-66151-9_1.

Gamez, Patrick. 2022. "A Friendly Critique of Levinasian Machine Ethics." *Southern Journal of Philosophy*, March 22, 2022. https://doi.org/10.1111/sjp.12455.

Gellers, Joshua C. 2020. *Rights for Robots: Artificial Intelligence, Animal and Environmental Law.* New York: Routledge. https://doi.org/10.4324/9780429288159.

Gellers, Joshua C., and David J. Gunkel. 2022. "Artificial Intelligence and International Human Rights Law: Implications for Humans and Technology in the 21st Century and Beyond." In *Handbook on the Politics and Governance of Big Data and Artificial Intelligence*, edited by Andrej Zwitter and Oscar J. Gstrein. Cheltenham: Edward Elgar.

Gerdes, Anne. 2015. "The Issue of Moral Consideration in Robot Ethics." *ACM SIGCAS Computers & Society* 45 (3): 274–280. https://doi.org/10.1145/2874239.2874278.

Gerdes, Anne, and Peter Øhrstrøm. 2015. "Issues in Robot Ethics Seen through the Lens of a Moral Turing Test." *Journal of Information, Communication and Ethics in Society* 13 (2): 98–109. https://doi.org/10.1108/JICES-09-2014-0038.

Ginway, M. Elizabeth. 2011. *Brazilian Science Fiction: Cultural Myths and Nationhood in the Land of the Future.* Lewisburg, PA: Bucknell University Press.

Gladden, Matthew E. 2016. "The Diffuse Intelligent Other: An Ontology of Nonlocalizable Robots as Moral and Legal Actors." In *Social Robots: Boundaries, Potential, Challenges*, edited by Marco Nørskov, 177–198. Burlington, VT: Ashgate.

Glenn, Linda Macdonald. 2003. "Biotechnology at the Margins of Personhood: An Evolving Legal Paradigm." *Journal of Evolution and Technology* 13 (October). http://jetpress.org/volume13/glenn.html.

Göcke, Benedikt Paul. 2020. "Could Artificial General Intelligence Be an End-in-Itself?" In *Artificial Intelligence: Reflections in Philosophy, Theology, and the Social Sciences*, edited by Benedikt Paul Göcke and Astrid Rosenthal-von der Pütten, 221–240. Leiden: Brill/Mentis Verlag.

Goertzel, Ben. 2002. "Thoughts on AI Morality." *Dynamical Psychology: An International, Interdisciplinary Journal of Complex Mental Processes*, May 2002. http://www.goertzel.org/dynapsyc/2002/AIMorality.htm.

Goldberg, Steven. 1996. "The Changing Face of Death: Computers, Consciousness, and Nancy Cruzan." *Stanford Law Review*, no. 43, 659–684. https://doi.org/10.2307/1228915.

Gordon, John-Stewart. 2021a. "Artificial Moral and Legal Personhood." *AI & Society* 36: 457–471. https://doi.org/10.1007/s00146-020-01063-2.

Gordon, John-Stewart. 2021b. "What Do We Owe to Intelligent Robots?" In *Smart Technologies and Fundamental Rights*, edited by John-Stewart Gordon, 17–47. Leiden: Brill/Rodopi.

Gordon, John-Stewart, and Ausrine Pasvenskiene. 2021. "Human Rights for Robots? A Literature Review." *AI and Ethics*, no. 1, 579–591. https://doi.org/10.1007/s43681-021-00050-7.

Gottlieb, Paula. 2019. "Aristotle on Non-contradiction." In *The Stanford Encyclopedia of Philosophy*, edited by Edward N. Zalta, Spring 2019 Edition. https://plato.stanford.edu/entries/aristotle-noncontradiction/.

Graness, Anke. 2018. "Becoming a Person: Personhood and Its Preconditions." In *Ubuntu and Personhood*, edited by James Ogude, 39–60. London: Africa World Press.

Grimm, Cindy M., and Kristen Thomasen. 2021. "On the Practicalities of Robots in Public Spaces." Paper presented at We Robot, University of Miami School of Law, Miami, FL, September 23–25. https://werobot2021.com/wp-content/uploads/2021/08/GrimmThomasen_Sidewalk-Robots.pdf.

Gunkel, David J. 1997. "Scary Monsters: Hegel and the Nature of the Monstrous." *International Studies in Philosophy* 29 (2): 24–46. https://doi.org/10.5840/intstudphil199729232.

Gunkel, David J. 2007. *Thinking Otherwise: Philosophy, Communication, Technology.* West Lafayette, IN: Purdue University Press.

Gunkel, David J. 2012. *The Machine Question: Critical Perspectives on AI, Robots and Ethics.* Cambridge, MA: MIT Press.

Gunkel, David. J. 2014. "A Vindication of the Rights of Machines." *Philosophy & Technology* 27 (1): 113–132. https://doi.org/10.1007/s13347-013-0121-z.

Gunkel, David J. 2016. "Of Remixology: Ethics and Aesthetics after Remix." Cambridge, MA: MIT Press.

Gunkel, David J. 2018. *Robot Rights.* Cambridge, MA: MIT Press.

Gunkel, David J. 2020a. *How to Survive a Robot Invasion: Rights, Responsibility, and AI.* New York: Routledge.

Gunkel, David J. 2020b. "Mind the Gap: Responsible Robotics and the Problem of Responsibility." *Ethics and Information Technology* 22:307–320. https://doi.org/10.1007/s10676-017-9428-2.

Gunkel, David J. 2021. *Deconstruction.* Cambridge, MA: MIT Press.

Gunkel, David J., Ciro Marcondes Filho, and Deiter Mersch. 2016. *The Changing Face of Alterity: Communication, Technology, and Other Subjects.* London: Rowman Littlefield.

Gunkel, David J., and Jordan Wales. 2021. "Debate: What Is Personhood in the Age of AI?" *AI & Society* 36:473–486. https://doi.org/10.1007/s00146-020-01129-1.

Habermas, Jürgen. 1987. *The Philosophical Discourse of Modernity: Twelve Lectures.* Translated by Frederick Lawrence. Cambridge, MA: MIT Press.

Hall, J. Storrs. 2001. "Ethics for Machines." Kurzweil, July 5, 2001. http://www.kurzweilai.net/ethics-for-machines.

Hamer, Richard, Lauren John, and Alexandra Moloney. 2021. "World First: Australia Says 'Yes' to AI Inventors." *Allens*, July 30, 2021. https://www.allens.com.au/insights-news/insights/2021/07/world-first-australia-says-yes-to-ai-inventors/.

Hampton, Gregory Jerome. 2015. *Imagining Slaves and Robots in Literature, Film, and Popular Culture: Reinventing Yesterday's Slave with Tomorrow's Robot.* New York: Lexington Books.

Harari, Yuval. 2015. *Sapiens: A Brief History of Humankind.* London: Random House.

Haraway, Donna J. 1991. *Simians, Cyborgs, and Women: The Reinvention of Nature.* New York: Routledge.

Haraway, Donna J. 2008. *When Species Meet.* Minneapolis: University of Minnesota Press.

Harman, Graham. 2002. *Tool Being: Heidegger and the Metaphysics of Objects*. Peru, IL: Open Court Publishing.

Harris, Jamie, and Jacy Reese Anthis. 2021. "The Moral Consideration of Artificial Entities: A Literature Review." *Science and Engineering Ethics* 27 (53): 1–95. https://doi.org/10.1007/s11948-021-00331-8.

Hart, Robert David. 2018. "Saudi Arabia's Robot Citizen Is Eroding Human Rights." *Quartz*, February 14, 2018. https://qz.com/1205017/saudi-arabias-robot-citizen-is-eroding-human-rights/.

Harvard Law Review. 2001. "Notes: What We Talk about When We Talk about Persons: The Language of a Legal Fiction." *Harvard Law Review* 114 (6): 1745–1768. https://doi.org/10.2307/1342652.

Hegel, G. W. F. 1969. *Enzyklopädie der Philosophischen Wissenschaften im Grundrisse*. Hamburg: Verlag von Felix Meiner. Translated by the author.

Hegel, G. W. F. 1977. *Phenomenology of Spirit*. Translated by A. V. Miller. Oxford: Oxford University Press. Originally published 1807.

Hegel, G. W. F. 2010. *The Science of Logic*. Translated by George Di Giovanni. Cambridge: Cambridge University Press. Originally published 1816.

Heidegger, Martin. 1962. *Being and Time*. Translated by John Macquarrie and Edward Robinson. New York: Harper & Row. Originally published 1927.

Heidegger, Martin. 1971. *Poetry, Language, Thought*. Trans. by Albert Hofstadter. New York: Harper and Row. Originally published 1954.

Heidegger, Martin. 1977. *The Question Concerning Technology and Other Essays*. Translated by W. Lovitt. New York: Harper & Row. Originally published 1962.

Heidegger, Martin. 2012. *Bremen and Freiburg Lectures: Insight into That Which Is and the Basic Principles of Thinking*. Translated by Andrew J. Mitchell. Bloomington: Indiana University Press. Originally published 1994.

Heller, Nathan. 2016. "If Animals Have Rights, Should Robots?" *New Yorker*, November 28, 2016. https://www.newyorker.com/magazine/2016/11/28/if-animals-have-rights-should-robots.

Higginbotham, A. Leon, Jr., and Barbara K. Kopytoff. 1989. "Property First, Humanity Second: The Recognition of the Slave's Human Nature in Virginia Civil Law." *Ohio State Law Journal*, no. 50, 511–540. https://heinonline.org/HOL/LandingPage?handle=hein.journals/ohslj50&div=28&id=&page=.

Hohfeld, Wesley. 1920. *Fundamental Legal Conceptions as Applied in Judicial Reasoning*. New Haven, CT: Yale University Press.

hooks, bell. 1992. *Black Looks: Race and Representation*. Boston: South End Press.

Hubbard, F. Patrick. 2011. "'Do Androids Dream?' Personhood and Intelligent Artifacts." *Temple Law Review* 83 (Winter): 405–474. https://scholarcommons.sc.edu /law_facpub/52/.

Hunt Botting, Eileen. 2021. *Artificial Life after Frankenstein*. Philadelphia: University of Pennsylvania Press.

Ihde, Don. 1990. *Technology and the Lifeworld: From Garden to Earth*. Bloomington: Indiana University Press.

Ihara, Craig K. 2004. "Are Individual Rights Necessary?" *In Confucian Ethics: A Comparative Study of Self, Autonomy, and Community*, edited by Kwong-Loi Shun and David B. Wong, 11–30. Cambridge: Cambridge University Press.

Inayatullah, Sohail, and Phil McNally. 1988. "The Rights of Robots: Technology, Culture and Law in the 21st Century." *Futures* 20 (2): 119–136. Also available at http://www.kurzweilai.net/the-rights-of-robots-technology-culture-and-law-in-the -21st-century.

IPWatchdog. 2021. "DABUS Gets Its First Patent in South Africa Under Formalities Examination." IP Watch Dog, July 29, 2021. https://www.ipwatchdog.com/2021 /07/29/dabus-gets-first-patent-south-africa-formalities-examination/id=136116/.

Iwai, Katsuhito. 1999. "Persons, Things and Corporations: The Corporate Personality Controversy and Comparative Corporate Governance." *American Journal of Comparative Law* 47 (4): 583–632. http://www.jstor.org/stable/841070.

Izumo, Takashi. 2018. "Digital Specific Property of Robots: A Historical Suggestion from Roman Law." *Delphi* 1 (1): 14–19. https://doi.org/10.21552/delphi/2018/1/8.

Jacobs, Harriet Ann. 2001. *Incidents in the Life of a Slave Girl*. New York: Dover. Originally published 1861.

Jaynes, Tyler L. 2020. "Legal Personhood for Artificial Intelligence: Citizenship as the Exception to the Rule." *AI & Society*, no. 35, 343–354. https://doi.org/10.1007 /s00146-019-00897-9.

Jecker, Nancy S. 2021. "Can We Wrong a Robot?" *AI & Society*. https://doi.org /10.1007/s00146-021-01278-x.

Jecker, Nancy S., Caesar A. Atiure, and Martin Odei Ajei. 2022. "The Moral Standing of Social Robots: Untapped Insights from Africa." *Philosophy & Technology* 35 (34): 1–22. https://doi.org/10.1007/s13347-022-00531-5.

Jibo. 2014. "JIBO: The World's First Social Robot for the Home." https://www .youtube.com/watch?v=H0h20jRA5M0.

Johnson, Barbara. 1987. *A World of Difference*. Baltimore: Johns Hopkins University Press.

Johnson, Brian David. 2011. *Science Fiction Prototyping: Designing the Future with Science Fiction*. Williston, VT: Morgan and Claypool. https://doi.org/10.2200/S00336 ED1V01Y201102CSL003.

Johnson, Deborah G. 2006. "Computer Systems: Moral Entities but Not Moral Agents." *Ethics and Information Technology*, no. 8, 195–204. https://doi.org/10.1007 /s10676-006-9111-5.

Jones, Raya. 2016. *Personhood and Social Robotics: A Psychological Consideration*. New York: Routledge.

Jordan, John. 2016. *Robots*. Cambridge, MA: MIT Press.

Jowitt, Joshua. 2021. "Assessing Contemporary Legislative Proposals for Their Compatibility with a Natural Law Case for AI Legal Personhood." *AI & Society*, no. 36, 499–508. https://doi.org/10.1007/s00146-020-00979-z.

Kant, Immanuel. 1965. *Critique of Pure Reason*. Translated by Norman Kemp Smith. New York: St. Martin's Press. Originally published 1781/1787.

Kant, Immanuel. 1985. *Critique of Practical Reason*. Translated by Lewis White Beck. New York: Macmillan. Originally published 1788.

Kant, Immanuel. 1987. *Critique of Judgment*. Translated by Werner S. Pluhar. Indianapolis, IN: Hackett Publishing Company. Originally published 1790.

Kant, Immanuel. 2012. *Groundwork of the Metaphysics of Morals*. Translated by Mary Gregor and Jens Timmermann. Cambridge: Cambridge University Press. Originally published 1786.

Kant, Immanuel. 2017. *The Metaphysics of Morals*. Translated by Mary Gregor. Cambridge: Cambridge University Press. Originally published 1797.

Kantayya, Shalini, dir. 2021. *Coded Bias*. 7th Empire Media.

Kaplan, Jerry. 2016. *Artificial Intelligence: What Everyone Needs to Know*. New York: Oxford University Press.

Karnow, Curtis E. A. 1996. "Liability for Distributed Artificial Intelligences." *Berkeley Technology Law Journal* 11 (1): 147–204. https://www.jstor.org/stable/24115584.

Katz, Andrew. 2008. "Intelligent Agents and Internet Commerce in Ancient Rome." *Computers and Law*. https://www.scl.org/articles/1095-intelligent-agents-and-internet -commerce-in-ancient-rome.

Katz, Andrew, and Michaela MacDonald. 2020. "Autonomous Intelligent Agents and the Roman Law of Slavery." In *Future Law: Emerging Technology, Regulation and Ethics*, edited by Lilian Edwards, Schafer Burkhard, and Edina Harbinja, 295–312. Edinburgh: Edinburgh University Press.

Kelsen, Hans. 1945. *General Theory of Law and State*. Translated by A. Wedberg. Cambridge, MA: Harvard University Press.

Kelsen, Hans. 1967. *Pure Theory of Law*. Translated by Max Knight. Berkeley: University of California Press.

Kerr, Ian R. 1999. "Providing for Autonomous Electronic Devices in the Uniform Electronic Commerce Act." Paper presented at Uniform Law Conference of Canada, Winnipeg, Manitoba, August 15–19.

Kierkegaard, Søren. 1987. *Either/Or, Part II*. Translated by Howard V. Hong and Edna H. Hong. Princeton, NJ: Princeton University Press.

Kim, Min-Sun. 2022. "Meta-narratives on Machinic Otherness: Beyond Anthropocentrism and Exoticism." *AI & Society*. https://doi.org/10.1007/s00146-022-01404-3.

Kim, Tae Wan. 2020. "Should Robots Have Rights or Rites? A Confucian Cross-Cultural Exploration to Robot Ethics." December 21, 2020: 1–23. Available at SSRN. http://dx.doi.org/10.2139/ssrn.3753070.

King, Edward. 2013. *Science Fiction and Digital Technologies in Argentine and Brazilian Culture*. New York: Palgrave Macmillan.

Kirchin, Simon. 2012. *Metaethics*. New York: Palgrave Macmillan.

Kitcher, Patricia. 1979. "Natural Kinds and Unnatural Persons." *Philosophy* 54 (210): 541–547. https://www.jstor.org/stable/3751047.

Knapp, Alex. 2011. "Should Artificial Intelligences Be Granted Civil Rights?" *Forbes*, April 4, 2011. https://www.forbes.com/sites/alexknapp/2011/04/04/should-artificial-intelligences-be-granted-civil-rights/.

Koops, Bert-Japp, Mireille Hildebrandt, and David-Oliver Jaquet-Chiffelle. 2010. "Bridging the Accountability Gap: Rights for New Entities in the Information Society?" *Minnesota Journal of Law, Science & Technology* 11 (2): 497–561. https://scholarship.law.umn.edu/mjlst/vol11/iss2/4.

Krämer, Carmen. 2020. "Can Robots Have Dignity?" In *Artificial Intelligence: Reflections in Philosophy, Theology, and the Social Sciences*, edited by Benedikt Paul Goecke and Astrid Rosenthal-von der Pütten, 241–253. Leiden, Netherlands: Brill. https://doi.org/10.30965/9783957437488_016.

Kramer, Matthew H., N. E. Simmonds, and Hillel Steiner, eds. 1998. *A Debate Over Rights: Philosophical Enquiries*. Oxford: Clarendon Press.

Kurki, Visa A. J. 2019. *A Theory of Legal Personhood*. Oxford: Oxford University Press.

Kurzgesagt—in a Nutshell. 2017. "Do Robots Deserve Rights?" Kurzgesagt—in a Nutshell. February 23, 2017. YouTube video, 6:34. https://www.youtube.com/watch?v=DHyUYg8X31c.

LaGrandeur, Kevin. 2013. *Androids and Intelligent Networks in Early Modern Literature and Culture.* New York: Routledge.

Latour, Bruno. 1993. *We Have Never Been Modern.* Translated by Catherine Porter. Cambridge, MA: Harvard University Press.

Lavender, Isiah. 2011. *Race in American Science Fiction.* Bloomington: Indiana University Press.

Lee, Keekok. 1999. *The Natural and the Artefactual: The Implications of Deep Science and Deep Technology for Environmental Philosophy.* New York: Lexington Books.

Lehman-Ludwig, Annie. 2019. "Robot Rights?" *Brown Political Review*, December 21, 2019. https://brownpoliticalreview.org/2019/12/robot-rights/.

Lehman-Wilzig, Sam N. 1981. "Frankenstein Unbound: Towards a Legal Definition of Artificial Intelligence." *Futures* 13 (6): 442–457. https://doi.org/10.1016/0016-3287 (81)90100-2.

Leiber, Justin. 1985. *Can Animals and Machines Be Persons? A Dialogue.* Indianapolis: Hackett.

Leopold, Aldo. 1966. *A Sand County Almanac.* New York: Oxford University Press.

Levinas, Emmanuel. 1969. *Totality and Infinity: An Essay on Exteriority.* Translated by A. Lingis. Pittsburgh: Duquesne University. Originally published 1961.

Levy, David. 2005. *Robots Unlimited: Life in a Virtual Age.* Boca Raton, FL: CRC Press.

Levy, David. 2009. "The Ethical Treatment of Artificially Conscious Robots." *International Journal of Social Robotics* 1 (3): 209–216. https://doi.org/10.1007/s12369 -009-0022-6.

Lewis, Jason Edward, Noelani Arista, Archer Pechawis, and Suzanne Kite. 2018. "Making Kin with the Machines." *Journal of Design and Science*, July 16, 2018. https:// doi.org/10.21428/bfafd97b.

Lima, Gabriel, Assem Zhunis, Lev Manovich, and Meeyoung Cha. 2021. "On the Social-Relational Moral Standing of AI: An Empirical Study Using AI-Generated Art." *Frontiers in Robotics and AI*, August 5, 2021. https://doi.org/10.3389/frobt.2021 .719944.

Lind, Douglas. 2009. "Pragmatism and Anthropomorphism: Reconceiving the Doctrine of the Personality of the Ship." *University of San Francisco Maritime Law Journal* 22 (1): 39–122.

Locke, John. 1996. *An Essay Concerning Human Understanding.* Edited by Kennith P. Winkler. Indianapolis: Hackett. Originally published 1689.

Loh, Janina. 2021. "Ascribing Rights to Robots as Potential Moral Patients." In *Smart Technologies and Fundamental Rights*, edited by John-Stewart Gordon, 101–126. Leiden: Brill Rodopi.

Lorde, Audre. 1984. *Sister Outsider: Essays and Speeches*. Berkeley, CA: Crossing Press.

Lyotard, Jean-François. 1984. *The Postmodern Condition: A Report on Knowledge*. Translated by Geoff Bennington and Brian Massumi. Minneapolis: University of Minnesota Press.

Marko, Kurt. 2019. "Robot Rights—a Legal Necessity or Ethical Absurdity?" Diginomica, January 3, 2019. https://diginomica.com/robot-rights-a-legal-necessity-or -ethical-absurdity.

Matthias, Andrew. 2004. "The Responsibility Gap: Ascribing Responsibility for the Actions of Learning Automata." *Ethics and Information Technology*, no. 6, 175–183. https://doi.org/10.1007/s10676-004-3422-1.

Mauss, Marcel. 1985. "A Category of the Human Mind: The Notion of Person; the Notion of Self." Translated by W. D. Halls. In *The Category of the Person*, edited by Michael Carrithers, Steven Collins, and Steven Lukes, 1–25. Cambridge: Cambridge University Press.

Marx, Johannes, and Christine Tiefensee. 2015. "Of Animals, Robots and Men." *Historical Social Research (Köln)* 40 (4): 70–91. https://doi.org/10.12759/hsr.40.2015.4.70-91.

Mayor, Adrienne. 2018. *Gods and Robots: Myths, Machines and Ancient Dreams of Technology*. Princeton, NJ: Princeton University Press.

Mbiti, John. 1969. *African Religions and Philosophy*. London: Heinemann Educational Books.

McLachlan, Hugh. 2019. "Ethics of AI: Should Sentient Robots Have the Same Rights as Humans?" *Independent*, June 26, 2019. https://www.independent.co.uk/news/science /ai-robots-human-rights-tech-science-ethics-a8965441.html.

McLuhan, Marshall, and Quentin Fiore. 2001. *War and Peace in the Global Village*. Berkeley, CA: Ginko Press.

Menkiti, Ifeanyi A. 1984. "Person and Community in African Traditional Thought." In *African Philosophy: An Introduction*, edited by Richard Wright, 171–182. Lanham, MD: University Press of America.

Miller, Lantz Fleming. 2015. "Granting Automata Human Rights: Challenge to a Basis of Full-Rights Privilege." *Human Rights Review* 16 (4): 369–391. https://doi.org /10.1007/s12142-015-0387-x.

Miller, Lantz Fleming. 2020. "Responsible Research for the Construction of Maximally Humanlike Automata: The Paradox of Unattainable Informed Consent. *Ethics and Information Technology* 22:297–305. https://doi.org/10.1007/s10676-017-9427-3.

Minsky, Marvin. 1994. "A Conversation with Marvin Minsky about Agents." *Communications of the ACM* 37 (7): 22–29. https://doi.org/10.1145/176789.176791.

Mocanu, Diana Mădălina. 2022. "Gradient Legal Personhood for AI Systems—Painting Continental Legal Shapes Made to Fit Analytical Molds." *Frontiers in Robotics and AI*, January 19, 2022. https://doi.org/10.3389/frobt.2021.788179.

Mosakas, Kęstutis. 2021a. "Machine Moral Standing: In Defense of the Standard Properties-Based View." In *Smart Technologies and Fundamental Rights*, edited by John-Stewart Gordon, 73–100. Leiden: Brill Rodopi.

Mosakas, Kęstutis. 2021b. "On the Moral Status of Social Robots: Considering the Consciousness Criterion." *AI & Society*, no. 36, 429–443. https://doi.org/10.1007/s00146-020-01002-1.

Moser, Paul K., and Thomas L. Carson. 2001. *Moral Relativism: A Reader.* New York: Oxford University Press.

Müller, Vincent C. 2021. "Is It Time for Robot Rights? Moral Status in Artificial Entities." *Ethics and Information Technology*, no. 23, 579–587. https://doi.org/10.1007/s10676-021-09596-w.

Muñoz, José Esteban, Jinthana Haritaworn, Myra Hird, Zakiyyah Iman Jackson, Jasbir K. Puar, Eileen Joy, Uri McMillan, Susan Stryker, Kim TallBear, Jami Weinstein, and Jack Halberstam. 2015. "Theorizing Queer Inhumanisms." *GLQ: A Journal of Lesbian and Gay Studies* 21 (2–3): 209–248. https://doi.org/10.1215/10642684-2843323.

Musiał, Maciej. 2017. "Designing (Artificial) People to Serve—the Other Side of the Coin." *Journal of Experimental & Theoretical Artificial Intelligence* 29 (5): 1087–1097. https://doi.org/10.1080/0952813X.2017.1309691.

Naffine, Ngaire. 2003. "Who Are Law's Persons? From Cheshire Cats to Responsible Subjects." *Modern Law Review* 66 (3): 346–367. https://doi.org/10.1111/1468-2230.6603002.

Naffine, Ngaire. 2009. *Law's Meaning of Life: Philosophy, Religion, Darwin and the Legal Person.* Oxford: Hart Publishing.

Natale, Simone. 2021. *Deceitful Media: Artificial Intelligence and Social Life after the Turing Test.* New York: Oxford University Press.

Navon, Mois. 2021. "The Virtuous Servant Owner—a Paradigm Whose Time Has Come (Again)." *Frontiers in Robotics and AI* 8 (715849): 1–15. https://doi.org/10.3389/frobt.2021.715849.

Neely, Erica L. 2012. "Machines and the Moral Community." In *The Machine Question: AI, Ethics and Moral Responsibility*, edited by Joanna Bryson, David J. Gunkel, and Steve Torrance, 38–42. Proceedings of AISB/IACAP World Congress 2012.

Neely, Erica L. 2014. "Machines and the Moral Community." *Philosophy & Technology* 27 (1): 97–111. https://doi.org/10.1007/s13347-013-0114-y.

Neuhäuser, Christian. 2015. "Some Skeptical Remarks Regarding Robot Responsibility and a Way Forward." In *Collective Agency and Cooperation in Natural and Artificial Systems: Explanation, Implementation, and Simulation*, edited by Catrin Misselhorn, 131–148. New York: Springer.

New Zealand Parliament. 2017. Te Awa Tupua (Whanganui River Claims Settlement) Act 2017, section 14.

Nietzsche, Friedrich. 1974. *The Gay Science*. Translated by Walter Kaufmann. New York: Vintage Books. Originally published 1882/1887.

Nourbakhsh, Illah. 2013. *Robot Futures*. Cambridge: MIT Press.

Nyholm, Sven. 2020. *Humans and Robots: Ethics, Agency and Anthropomorphism*. New York: Rowman Littlefield.

Nyholm, Sven, and Lily E. Frank. 2017. "From Sex Robots to Love Robots: Is Mutual Love with a Robot Possible?" In *Robot Sex: Social and Ethical Implications*, edited by John Danaher and Neil McArthur, 219–244. Cambridge, MA: MIT Press.

Pagallo, Ugo. 2011. "Killers, Fridges, and Slaves: A Legal Journey in Robotics." *AI & Society* 26:347–354. https://doi.org/10.1007/s00146-010-0316-0.

Pagallo, Ugo. 2012. "Three Roads to Complexity, AI and the Law of Robots: On Crimes, Contracts and Torts." *In AI Approaches to the Complexity of Legal Systems—Models and Ethical Challenges for Legal Systems, Legal Language and Legal Ontologies, Argumentation and Software Agents*, edited by Monica Palmirani, Ugo Pagallo, Pompeu Casanovas and Giovanni Sartor, 48–60. Berlin, Springer.

Pagallo, Ugo. 2013. *The Laws of Robots: Crimes, Contracts, and Torts*. Dordrecht: Springer.

Pagallo, Ugo. 2018. "Vital, Sophia, and Co.—the Quest for the Legal Personhood of Robots." *Information* 9 (9): 1–11. https://doi.org/10.3390/info9090230.

Persaud, Priya, Aparna S. Varde, and Weitian Wang. 2021. "Can Robots Get Some Human Rights? A Cross-Disciplinary Discussion." *Journal of Robotics* 2021 (5461703). https://doi.org/10.1155/2021/5461703.

Petersen, Steve. 2007. "The Ethics of Robot Servitude." *Journal of Experimental & Theoretical Artificial Intelligence* 19 (1): 43–54. http://dx.doi.org/10.1080/0952813060 1116139.

Petersen, Steve. 2011. "Designing People to Serve." In *Robot Ethics: The Ethical and Social Implications of Robotics*, edited by Patrick Lin, Keith Abney, and George A. Bekey, 283–298. Cambridge, MA: MIT Press.

Pietrzykowski, Tomasz. 2018. *Personhood beyond Humanism: Animals, Chimeras, Autonomous Agents and the Law*. Cham: Springer.

Plato. 1982. *Plato I: Euthyphro, Apology, Crito, Phaedo, Phaedrus*. Translated by Harold North Fowler. Cambridge, MA: Harvard University Press.

Plato. 1987. *Plato VII: Theaetetus, Sophist*. Translated by H. N. Fowler. Cambridge, MA: Harvard University Press.

Pottage, Alan. 2004. "Introduction: The Fabrication of Persons and Things." In *Law, Anthropology, and the Constitution of the Social: Making Persons and Things*, edited by Alan Pottage and Martha Mundy, 1–39. Cambridge: Cambridge University Press.

Puig de la Bellacasa, María. 2017. *Matters of Care: Speculative Ethics in More than Human Worlds*. Minneapolis: University of Minnesota Press.

Putnam, Hilary. 1964. "Robots: Machines or Artificially Created Life?" *Journal of Philosophy* 61 (21): 668–691. http://www.jstor.org/stable/2023045.

Rademeyer, Leon B. 2017. "Legal Rights for Robots by 2060?" *Knowledge Futures: Interdisciplinary Journal of Futures Studies* 1 (1). https://research.usc.edu.au/esploro /outputs/99451189902621.

Rancière, Jacques, Davide Panagia, and Rachel Bowlby. 2001. "Ten Theses on Politics." *Theory & Event* 5 (3). https://doi.org/10.1353/tae.2001.0028.

Reeves, Byron, and Clifford Nass. 1996. *The Media Equation: How People Treat Computers, Television, and New Media Like Real People and Places*. Cambridge: Cambridge University Press.

Regan, Tom. 1983. *The Case for Animal Rights*. Berkeley: University of California Press.

Reiss, Michael J. 2021. "Robots as Persons? Implications for Moral Education." *Journal of Moral Education* 50 (1): 68–76. https://doi.org/10.1080/03057240.2020.1763933.

Revolidis, Ioannis, and Alan Dahi. 2018. "The Peculiar Case of the Mushroom Picking Robot: Extra-contractual Liability in Robotics." In *Robotics, AI and the Future of Law*, edited by Marcelo Corrales, Mark Fenwick, and Nikolaus Forgó, 57–79. Singapore: Springer.

Richardson, Kathleen. 2019. "The Human Relationship in the Ethics of Robotics: A Call to Martin Buber's *I and Thou*." *AI & Society* 34 (1): 75–82. https://doi.org /10.1007/s00146-017-0699-2.

Ricœur, Paul. 2007. *Reflections on the Just*. Translated by David Pellauer. Chicago: University of Chicago Press.

Risely, James. 2016. "Microsoft's Millennial Chatbot Tay.ai Pulled Offline after Internet Teaches Her Racism." *GeekWire*, March 24, 2016. http://www.geekwire.com/2016 /even-robot-teens-impressionable-microsofts-tay-ai-pulled-internet-teaches-racism/.

Rivard, Michael D. 1992. "Toward a General Theory of Constitutional Personhood: A Theory of Constitutional Personhood for Transgenic Humanoid Species." *UCLA Law Review* 39 (5): 1425–1510.

Robertson, Jennifer. 2014. "Human Rights vs. Robot Rights: Forecasts from Japan." *Critical Asian Studies* 46 (4): 571–598. https://doi.org/10.1080/14672715.2014.960707.

Roh, Daniel. 2009. "Do Humanlike Machines Deserve Human Rights?" *Wired*, January 19, 2009. https://www.wired.com/2009/01/st-essay-16.

Rorty, Amélie Oksenberg. 1976. "A Literary Postscript: Characters, Persons, Selves, Individuals." In *The Identities of Persons*, edited by Amélie Oksenberg Rorty, 301–324. Berkeley: University of California Press.

Rorty, Amélie Oksenberg. 1988. "Persons and Personae." In *Mind in Action*, edited by A. O. Rorty, 27–98. New York: Beacon Press.

Royle, Nicholas. 2009. *In Memory of Jacques Derrida*. Edinburgh: Edinburgh University Press.

Saerens, Patrick. 2020. "Le Droit des Robots, un Droit de l'Homme en Devenir?" *Communication, technologies et Développement*, no. 8, 1–8. https://doi.org/10.4000/ctd.2877.

Sætra, Henrik Skaug. 2021. "Challenging the Neo-anthropocentric Relational Approach to Robot Rights." *Frontiers in Robotics and AI* 8 (744426): 1–9. https://doi.org/10.3389/frobt.2021.744426.

Sætra, Henrik Skaug, and Eduard Fosch-Villaronga. 2021. "Research in AI Has Implications for Society: How Do We Respond?" *Morals & Machines* 1 (1): 60–73. https://doi.org/10.5771/2747-5174-2021-1-60.

Said, Edward W. 1979. *Orientalism*. New York: Vintage Books.

Salmond, John. 1907. *Jurisprudence, or the Legal Theory of Law*. London: Stevens and Haynes.

Samuel, Sigel. 2019. "Humans Keep Directing Abuse—Even Racism—at Robots." Vox, August 2, 2019. https://www.vox.com/future-perfect/2019/8/2/20746236/ai-robot-empathy-ethics-racism-gender-bias.

Sartor, G. 2002. "Agents in Cyberlaw." Paper presented at the Workshop on the Law of Electronic Agents (LEA 2002), Bologna, Italy, July 13.

Scheerer, Robert, dir. 1989. *Star Trek: The Next Generation*. Season 2, episode 9, "Measure of a Man." Aired February 11, 1989.

Schirmer, Jan-Erik. 2020. "Artificial Intelligence and Legal Personality: Introducing 'Teilrechtsfähigkeit': A Partial Legal Status Made in Germany." In *Regulating Artificial*

Intelligence, edited by Thomas Wischmeyer and Timo Rademacher, 123–142. Cham: Springer. https://doi.org/10.1007/978-3-030-32361-5_6.

Schröder, Wolfgang M. 2021. "Robots and Rights: Reviewing Recent Positions in Legal Philosophy and Ethics." In *Robotics, AI, and Humanity: Science, Ethics and Policy*, edited by Joachim von Braun, Margaret S. Archer, Gregory M. Reichberg, and Marcelo Sánchez Sorondo, 191–203. Cham: Springer.

Schwitzgebel, Eric, and Mara Garza. 2015. "A Defense of the Rights of Artificial Intelligences." *Midwest Studies in Philosophy* 39 (1): 89–119. https://doi.org/10.1111/misp.12032.

Schwitzgebel, Eric, and Mara Garza. 2020. "Designing AI with Rights, Consciousness, Self-Respect, and Freedom." In *Ethics of Artificial Intelligence*, edited by S. Matthew Liao. Oxford: Oxford University Press. https://doi.org/10.1093/oso/9780190905033.003.0017.

Scott, Robert L. 1976. "On Viewing Rhetoric as Epistemic: Ten Years Later." *Central States Speech Journal* 27 (4): 258–266. https://doi.org/10.1080/10510977609367902.

Searle, John. 1980. "Minds, Brains and Programs." *Behavioral and Brain Sciences* 3 (3): 417–457. https://doi.org/10.1017/S0140525X00005756.

Searle, John. 1999. "The Chinese Room." In *The MIT Encyclopedia of the Cognitive Sciences*, edited by R. A. Wilson and F. Keil, 115–116. Cambridge, MA: MIT Press.

Seuss, Dr. 1982. *Horton Hears a Who*. New York: Random House. Originally published 1954.

Shannon, Claude E. 1950. "Programming a Computer for Playing Chess." *Philosophical Magazine*, 7th ser., 41 (314): 256–275. https://doi.org/10.1080/14786445008521796.

Shaviro, Steven. 2016. *The Universe of Things: On Speculative Realism*. Minneapolis: University of Minnesota Press.

Sigfusson, Lauren. 2017. "Do Robots Deserve Human Rights?" *Discover*, December 5. https://www.discovermagazine.com/technology/do-robots-deserve-human-rights.

Singer, Peter. 1975. *Animal Liberation: A New Ethics for Our Treatment of Animals*. New York: New York Review of Books.

Singer, Peter, and Agata Sagan. 2009a. "Rights for Robots." *Project Syndicate*. https://www.project-syndicate.org/commentary/rights-for-robots.

Singer, Peter, and Agata Sagan. 2009b. "When Robots Have Feelings." *The Guardian*, December 14, 2009. https://www.theguardian.com/commentisfree/2009/dec/14/rage-against-machines-robots.Sloterdjik, Peter. 2001. *Nicht Gerettet: Versuche nach Heidegger*. Frankfurt am Main: Surkamp.

Smids, Jille. 2020. "Danaher's Ethical Behaviourism: An Adequate Guide to Assessing the Moral Status of a Robot?" *Science and Engineering Ethics*, no. 26, 2849–2866. https://doi.org/10.1007/s11948-020-00230-4.

Smith, Christian. 2010. *What Is a Person? Rethinking Humanity, Social Life, and the Moral Good from the Person Up.* Chicago: University of Chicago Press.

Smith, Joshua K. 2021a. *Robotic Persons: Our Future with Social Robots.* Bloomington, IN: WestBow.

Smith, Joshua K. 2021b. *Robot Theology: Old Questions through New Media.* Eugene, OR: Wipf and Stock Publishers.

Solum, Lawrence B. 1992. "Legal Personhood for Artificial Intelligences." *North Carolina Law Review* 70 (4): 1231–1287. http://scholarship.law.unc.edu/nclr/vol70/iss4/4.

Spaemann, Robert. 2006. *Persons: The Difference Between "Someone" and "Something".* Translated by Oliver O'Donovan. New York: Oxford University Press.

Sparrow, Robert. 2004. "The Turing Triage Test." *Ethics and Information Technology* 6 (4): 203–213. https://doi.org/10.1007/s10676-004-6491-2.

Sparrow, Robert. 2017. "Robots, Rape, and Representation." *International Journal of Social Robotics* 9 (4): 465–477. https://doi.org/10.1007/s12369-017-0413-z.

Stone, Christopher D. 1974. *Should Trees Have Standing? Toward Legal Rights for Natural Objects.* Los Altos, CA: William Kaufmann.

Strawson, Peter Frederick. 1959. *Individuals: An Essay in Descriptive Metaphysics.* London: Methuen & Co.

Stryker, Susan. 1994. "My Words to Victor Frankenstein above the Village of Chamounix: Performing Transgender Rage." *GLQ: A Journal of Lesbian and Gay Studies* 1 (3): 237–254. https://doi.org/10.1215/10642684-1-3-237.

Szollosy, Michael. 2017. "EPSRC Principles of Robotics: Defending an Obsolete Human(ism)?" *Connection Science* 29 (2): 150–159. https://doi.org/10.1080/09540091.2017.1279126.

Tasioulas, John. 2019. "First Steps towards an Ethics of Robots and Artificial Intelligence." *Journal of Practical Ethics* 7 (1): 49–83. http://www.jpe.ox.ac.uk/wp-content/uploads/2019/06/Tasioulas.pdf.

Taylor, Charles. 1985. "The Person." In *The Category of the Person*, edited by Michael Carrithers, Steven Collins, and Steven Lukes, 257–281. Cambridge: Cambridge University Press.

Taylor, Mark C. 1997. *Hiding.* Chicago: University of Chicago Press.

Taylor, Thomas. 1966. *A Vindication of the Rights of Brutes*. Gainesville, FL: Scholars' Facsimiles & Reprints. Originally published 1792.

Teichman, Jenny. 1985. "The Definition of Person." *Philosophy* 60 (232): 175–185. http://www.jstor.org/stable/3750997.

Teubner, Gunther. 2006. "Rights of Non-humans? Electronic Agents and Animals as New Actors in Politics and Law." *Journal of Law and Society* 33 (4): 497–521. https://www.jstor.org/stable/4129589.

Thibaut, Anton Friedrich Justus. 1855. *An Introduction to the Study of Jurisprudence*. Translated by Nathaniel Lindley. Philadelphia: T. & J. W. Johnson, Law Booksellers. Originally published 1814.

Tocqueville, Alexis de. 1899. *Democracy in America*. Translated by H. Reeve. New York: Colonial Press.

Tollon, Fabio. 2021. "The Artificial View: Toward a Non-anthropocentric Account of Moral Patiency." *Ethics and Information Technology* 23 (1): 147–155. https://doi.org/10.1007/s10676-020-09540-4.

Torrance, Steve. 2008. "Ethics and Consciousness in Artificial Agents." *AI & Society*, no. 22, 495–521. https://doi.org/10.1007/s00146-007-0091-8.

Torrance, Steve. 2014. "Artificial Consciousness and Artificial Ethics: Between Realism and Social Relationism." *Philosophy & Technology* 27 (1): 9–29. https://doi.org/10.1007/s13347-013-0136-5.

Trahan, J.-R. 2008. "The Distinction between Persons and Things: An Historical Perspective." *Journal of Civil Law Studies* 1 (1): 9–20. https://digitalcommons.law.lsu.edu/jcls/vol1/iss1/3.

Turing, Alan M. 1950. "Computing Machinery and Intelligence." *Mind* 59 (236): 433–460. https://doi.org/10.1093/mind/LIX.236.433. Reprinted in *Computer Media and Communication*, edited by Paul A. Mayer, 37–58. Oxford: Oxford University Press.

Turner, Jacob. 2019. *Robot Rules: Regulating Artificial Intelligence*. Cham: Palgrave Macmillan.

United States Supreme Court. 1819. Dartmouth College v. Woodward. 17 U.S. (4 Wheat.) 518. https://tile.loc.gov/storage-services/service/ll/usrep/usrep017/usrep017518/usrep017518.pdf.

van den Hoven van Genderen, Robert. 2018. "Do We Need New Legal Personhood in the Age of Robots and AI?" In *Robotics, AI and the Future of Law*, edited by Marcelo Corrales, Mark Fenwick, and Nikolaus Forgó, 15–55. Singapore: Springer. https://doi.org/10.1007/978-981-13-2874-9_2.

Velmans, Max. 2000. *Understanding Consciousness*. New York: Routledge.

Vincent, James. 2017. "Pretending to Give a Robot Citizenship Helps No One." The Verge, October 30, 2017. https://www.theverge.com/2017/10/30/16552006/robot -rights-citizenship-saudi-arabia-sophia.

Viveiros de Castro, Eduardo. 2015. *The Relative Native: Essays on Indigenous Conceptual Worlds*. Translated by Martin Holbraad, David Rodgers, and Julia Sauma. Chicago: HAU Press.

Viveiros de Castro, Eduardo. 2017. *Cannibal Metaphysics: For a Post-structural Anthropology*. Translated by Peter Skafish. Minneapolis: University of Minnesota Press.

Vogel, Steven. 2015. *Thinking like a Mall: Environmental Philosophy after the End of Nature*. Cambridge, MA: MIT Press.

Wachowski, Lana, and Lilly Wachowski, dirs. 1999. *The Matrix*. Burbank, CA: Warner Home Video.

Wagner, Gerhard. 2019. "Robot, Inc.: Personhood for Autonomous Systems?" *Fordham Law Review* 88 (2): 591–612. https://ir.lawnet.fordham.edu/flr/vol88/iss2/8.

Wakabayashi, Daisuke. 2018. "Self-Driving Uber Car Kills Pedestrian in Arizona, where Robots Roam." *New York Times*, March 19, 2018. https://www.nytimes.com/2018 /03/19/technology/uber-driverless-fatality.html.

Watson, A. 1988. *Roman Slave Law*. Baltimore: Johns Hopkins University Press.

Wenar, Leif. 2020. "Rights." In *The Stanford Encyclopedia of Philosophy*, edited by Edward N. Zalta, Spring 2021 Edition. https://plato.stanford.edu/archives/spr2021 /entries/rights/.

Wettig, Steffen, and Eberhard Zehendner. 2004. "A Legal Analysis of Human and Electronic Agents." *Artificial Intelligence and Law* 12 (1): 111–135. https://doi.org /10.1007/s10506-004-0815-8.

Willick, Marshall S. 1983. "Artificial Intelligence: Some Legal Approaches and Implications." *AI Magazine* 4 (2): 5–16. https://doi.org/10.1609/aimag.v4i2.392.

Willick, Marshall S. 1985. "Constitutional Law and Artificial Intelligence: The Potential Legal Recognition of Computers as 'Persons.'" Paper presented at the Ninth International Joint Conference on Artificial Intelligence, Los Angeles, CA. https:// www.ijcai.org/Proceedings/85-2/Papers/115.pdf.

Wiener, Norbert. 1988. *The Human Use of Human Beings: Cybernetics and Society*. Boston: Da Capo Press.

Wiener, Norbert. 1996. *Cybernetics: Or Control and Communication in the Animal and the Machine*. Cambridge, MA: MIT Press.

Winner, Langdon. 1977. *Autonomous Technology: Technics-out-of-Control as a Theme in Political Thought*. Cambridge, MA: MIT Press.

Wittgenstein, Ludwig. 1995. *Tractatus Logico-Philosophicus*. New York: Routledge. Originally published 1922.

Wojtczak, Sylwia. 2022. "Endowing Artificial Intelligence with Legal Subjectivity." *AI & Society*, no. 37, 205–213. https://doi.org/10.1007/s00146-021-01147-7.

Wollstonecraft, Mary. 1996. *A Vindication of the Rights of Men*. New York: Prometheus Books. Originally published 1790.

Wollstonecraft, Mary. 2004. *A Vindication of the Rights of Woman*. New York: Penguin Classics. Originally published 1792.

Wurah, Amanda. 2017. "We Hold These Truths to Be Self-Evident, that All Robots Are Created Equal." *Journal of Futures Studies* 22 (2): 61–74. https://doi.org/10.6531 /JFS.2017.22(2).A61.

Wright, R. George. 2019. "The Constitutional Rights of Advanced Robots (and of Human Beings)." *Arkansas Law Review* 71 (3): 613–646. https://scholarworks.uark .edu/alr/vol71/iss3/2.

Yolgormez, Ceyda, and Joseph Thibodeau. 2021. "Socially Robotic: Making Useless Machines." *AI & Society*, no. 37, 565–578. https://doi.org/10.1007/s00146-021 -01213-0.

Zimmerman, Michael. 1990. *Heidegger's Confrontation with Modernity: Technology, Politics, Art*. Bloomington: Indiana University Press.

Žižek, Slavoj. 2006. "Philosophy, the 'Unknown Knowns,' and the Public Use of Reason." *Topoi*, no. 25, 137–142. https://doi.org/10.1007/s11245-006-0021-2.

Žižek, Slavoj. 2008. *In Defense of Lost Causes*. London: Verso.

Žižek, Slavoj. 2016. *Disparities*. London: Bloomsbury Academic.

Index